AF575279

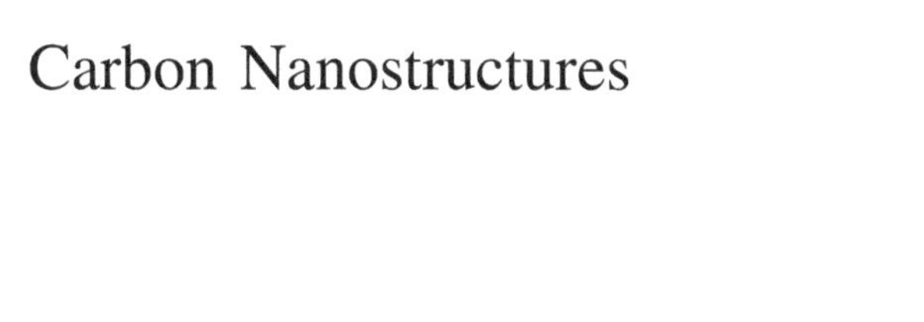

Carbon Nanostructures

For further volumes:
http://www.springer.com/series/8633

Oleksandr Loboda

Quantum-Chemical Studies on Porphyrins, Fullerenes and Carbon Nanostructures

Springer

Oleksandr Loboda
Institute of Colloid and Water Chemistry
Kiev
Ukraine

ISSN 2191-3005 ISSN 2191-3013 (electronic)
ISBN 978-3-642-31844-3 ISBN 978-3-642-31845-0 (eBook)
DOI 10.1007/978-3-642-31845-0
Springer Heidelberg New York Dordrecht London

Library of Congress Control Number: 2012943354

Printed on acid-free paper

Springer is part of Springer Science+Business Media (www.springer.com)

Acknowledgments

I want to express my sincere gratitude to my collaborators and coauthors Prof. H. Ågren, Prof. V. R. Jensen, Prof. M. G. Papadopoulos, and Prof. Y. Aoki for fruitful collaboration and interesting discussions.

I wish to thank my wife Natalia and daughter Alice, my sister's family, and my parents for their love and support.

Contents

Abbreviations

BDE	Bond dissociation energy
BSSE	Basis superposition error
CMO	Canonical molecular orbital
CPU	Central processing unit
CDA	Charge decomposition analysis
ECP	Effective core potential
EM	Elongation method
H_2TPP	Free base tetraphenyl porphion
HOMO	Highest occupied molecular orbital
HFC	Hyperfine coupling
LRC	Long-range corrected
LUMO	Lowest occupied molecular orbital
NLO	Nonlinear optics
NR	Nuclear relaxation
OLED	Organic light-emitting diode
$O(N)$	Linear scaling
OTE	Oligo(2,5-thienylene-ethynylene)
QR	Quadratic response
QFMM	Quantum fast multipole method
RLMO	Region localized molecular orbital
SOC	Spin–orbit coupling
SSC	Spin–spin coupling
SW-BN/CNT	Single wall boron nitride/carbon nanotube
TD DFT	Time-dependent density functional theory
TPA	Two-photon absorption
ZFS	Zero-field splitting

Abstract

This book contains a theoretical study of electronic structure, optical and spectroscopic properties of a number of compounds, in particular, porphyrins, fullerenes, and carbon nanostructures, which include heteroatom single wall nanotubes and others. The milestone of the book is to study the nonlinear optical properties and disclose the influence of the nature of substitutes to enhance nonlinear optical effects. The book presents qualitatively new results that address the issue that is of practical importance—namely, the discovery and development of materials for photosensitizers, organic LEDs, and solar cells. The performed analysis of a wide range of quantum-chemical methods aimed to study photophysical properties of porphyrin and fullerene derivatives. In particular, as a result of the work it was found the cause and relationship between the overshooted values of polarization components of the electron and the donor ability of the substitute. The electronic and vibrational contributions to the hyperpolarization of fullerene-chromophore donor–acceptor pairs are qualitatively and quantitatively evaluated. Single and two-photon absorption spectra of functionalized fullerene derivatives that were calculated on the basis of quantum-chemical calculations agree well with experimental results obtained by MALDI-TOF-MS, ATR-IR, UV, NMR, Z-scan spectroscopies. Particular attention is paid to the design and development of new $(O)N$ linear scaling methods for application in quantum-chemical theory of electronic structure. As shown in the book these methods display accuracy comparable to the accuracy of traditional method calculations; moreover, they are greatly reducing cost of CPU time, which make them an indispensable tool for studying the electronic structure of nano sized compounds.

Introduction

Actuality of the topic: The current importance of the work is in urgent need to use energy-saving technologies—namely, the creation of highly efficient materials for energy conversion of sunlight with a high quantum yield, photosensitizers, photoconductors, devices based on organic light-emitting diodes in computer and telecommunication networks. Most current methods for quantum-chemical calculations that can be applied to polyatomic systems have a whole series of approximations, which cannot preserve the high quality of results. However, for numerical integration, which is used to study NLO must have accuracy in the values of total energies at least up to the tenth decimal. Thus, the present book gives a good feedback to the increasing demands in developing effective precision linear quantum calculation methods for the study of linear and nonlinear optical properties of periodic and aperiodic polymeric structures with delocalized π system.

Goals and objectives of the study. Objective: to establish a relationship between the optical and magnetic properties, systematic scientific analysis of the nature of the interaction of metals with fullerenes, the study of nonlinear optical effects of porphyrins, fullerenes and carbon nanostructures.

To achieve these goals the following scientific and practical tasks should be performed:

- Optimization of geometrical structure: porphyrins, C_{60} derivatives, exo- and endohedral metallofullerenes, fullerene-porphyrin dyads, polyacenes, carbon nanotubes;
- Conduct a comprehensive charge decomposition analysis of metal atoms with fullerene compounds;
- Explore the phenomenon of adsorption of metals on the surface of fullerenes;
- Investigate the linear and nonlinear optical effects of porphyrin derivatives, fullerenes, and carbon nanostructures;
- Assess the impact of the nature of substituents on the optical properties of fullerene hybrid materials;

- Determine the cause of the broadening of line in optical detection of magnetic resonance for the triplet excited states of porphyrins;
- Develop effective methods for the investigation of the electronic structure and nonlinear optical properties of polyatomic molecular systems.

Object of research: fullerenes, porphyrins, carbon nanostructures.

Subject of research: electronic structure, optical, and magnetic properties.

Research methods: ab initio methods, density functional theory methods, semiempirical methods, O (N) methods for linear calculations (QFMM, EM).

Scientific novelty of the results: The most significant new scientific results of the study are established the linear and nonlinear optical (NLO) properties of porphyrins derivatives and fullerene carbon nanostructures.

The radiative lifetime of phosphorescence lines and microwave signals in optical detection of magnetic resonance (ODMR) spectra are obtained using the B3LYP hybrid density functional and the quadratic response method. The zero-field splitting (ZFS) in the lowest triplet state, a^3B_{2u}, of FBP is calculated as an expectation value of spin–spin coupling operator using the self-consistent field wave function. The second-order contribution to ZFS from the spin–orbit coupling operator is found to be almost negligible. The interpretation of the ODMR spectrum is completed by computing the hyperfine tensors of the ^{14}N, ^{13}C, and hydrogen atoms in the lowest triplet state. The most intensive phosphorescence emission corresponds to the T^z-spin sublevel of the a^3B_{2u} state, where the z-axis lies in the N–H direction of the FBP molecule in a qualitative agreement with ODMR data. The results indicate that the observed decay of the lowest triplet state of FBP molecule is determined by nonradiative deactivation. The calculated radiative rate constant for the T^z spin sublevel $k_z = 2.65 \times 10^{-3}\,s^{-1}$.

The performed study of photoinduced isomerisation reaction of free base porphin molecule using the DFT-B3LYP method show that the reaction occurs in a stepwise pathway of isomerisation.

A novel $Pd_2(\eta^2 - C_{60})$ structure with the two metal atoms bridging over a six-membered ring has been identified as the most stable arrangement of two palladium atoms on the surface of C_{60}. Both metal atoms benefit from η^2 coordination at (6–6′) junctions as well as some metal–metal interaction. Binding of Pd atoms to the fullerene is preferred over palladium dimerization.

Using a wide range of quantum-chemical methods the linear and nonlinear optical properties of [60]fullerene-chromophore dyads of different electron–donor character were analyzed in detail. It is demonstrated that in the case of investigated systems, traditional functionals poorly reproduce the values of first-order hyperpolarizability (β) calculated using MP2 method. It turned out, however, that both Coulomb-attenuating model (CAM–B3LYP) and LC–BLYP functional give reliable electronic contributions to β. The analysis of the relations between the nature of the chromophore and the properties of the whole [60]fullerene-chromophore dyad shows that substitution of 2,1,3-benzothiadiazole by triphenylamine group lead to significantly larger values of first- and second-order hyperpolarizability.

The two-photon absorption cross section is also enhanced upon chromophore modification. For [60]fullerene-chromophore dyads the harmonic vibrational contributions to β have been observed to be much larger than the electronic counterpart. The calculations of vibrational contributions to β for fulleropyrrolidine reveal, however, very large anharmonicity effects.

The linear and nonlinear optical properties of C60-triphenylamine (TPhA) hybrids are reported. The synthesized materials were prepared following the 1,3-dipolar cycloaddition of azomethine ylides onto the skeleton of C_{60} forming the TPhA-based monoadduct, equatorial bis-adduct and dumbell C_{60}. It was proved that in all considered cases, C_{60} serves as an acceptor while triphenylamine unit acts as a donor. It was found that the total second-order hyperpolarizability of C60-TphA-C60 system is several times larger than that of TPhA-C60. The results of experimental measurements are supported by quantum-chemical calculations.

Electronic and vibrational nuclear relaxation (NR) contributions to the dipole (hyper)polarizabilities of the endohedral fullerene $Li@C_{60}$ and its monovalent cation $[Li@C_{60}]^+$ have been carried out at the (U)B3LYP level. The obtained new results differ significantly from those reported previously using more approximate methods. The properties are compared with those of the corresponding hypothetical non-interacting systems with a valence electron transferred from Li to the cage. Whereas the NR contribution to the static linear polarizabilities is small in comparison with the corresponding electronic property, the opposite is true for the static hyperpolarizabilities. A relatively small, but non-negligible, NR contribution to the dc-Pockels effect is obtained in the infinite frequency approximation.

It is demonstrated that the functionalization of porphyrin by adding a metal atom significantly increases the NLO properties of the investigated systems. It was observed that increasing the size of the spacer chain from 0 to 8 carbon units almost linearly enhances the polarizability and second hyperpolarizability of [60]-H_2 P system.

A new approach has been developed for solving the eigenvalue problem for the oligomer chain systems based on localized molecular orbitals (LMO) of the separated fragments within the elongation method. The method performed in this work and implemented in elongation scheme yields excellent agreement with the conventional results. It has been demonstrated that the proposed algorithm for computing orbital energies and eigenvectors in elongation method reduces the CPU time usage up to 50 %.

Designed and implemented in software the new linear scaling technique for the study of compounds with delocalized electronic density.

The practical significance of the results: The research conducted in the dissertation is of great practical importance for the synthesis of optically active organic materials used in photonics to create a fundamentally new organic light-emitting diodes (LED) and to develop photoconductive and photosensitive elements for converting solar energy into electricity.

The developed linear scaling methods significantly reduce the CPU time of quantum studies for large multi-electron systems without significant loss of accuracy in total energy calculations compared to traditional methods of calculating the NLO.

Chapter 1
Porphyrins

1.1 Triplet State Properties of Free-Base Porphin

All life on the Earth depends on photosynthesis [1–3] which is a photochemical process in green plants and in some bacteria that has a very high quantum yield unusual for visible light energy transformation into organic chemical reactions. Important information about photosynthetic pigments has been obtained in studies of their luminescence, which is quite specific in comparison with typical organic dyes.

Free-base porphin (FBP) is the simplest macrocycle without any substitutes related to the porphyrinoide systems of chlorophylls, bacteriochlorophylls and other biological systems. In spite of the fundamental significance of FBP as a prototype of a large family of porphyrins even the basic photophysics of this molecule is incompletely understood [4]. Assignment of the symmetry of the lowest triplet state has only been made indirectly, on the basis of comparison of EPR spectra with the results of simple MO calculations [5–7]. The mechanism of the photochemical isomerisation of FBP and its derivatives (the photoinduced double hydrogen atom shift) [5, 8, 9] is still unclear. The radiative lifetime of FBP phosphorescence has only been indirectly estimated [6]. Like all porphyrins, FBP has two major absorption bands, namely the Q band in the visible region and the B band (or Soret band) in the near UV region, which have been the subject for many quantum chemical studies [4, 10–18]. In recent time there is a transfer from semiempirical calculations [10, 11] to more sophisticated approaches [4, 12, 13, 15–17]. For example, Nakatsuji et al. applied symmetry-adapted-cluster (SAC) and SAC-configuration interaction (CI) methods to FBP [12] and to a number of porphyrin-type molecules [13, 19] in order to interpret the singlet–singlet absorption spectrum. Roos et al. used their CASPT2 method for the same purpose [14]. Another promising approach is the time-dependent density functional theory (TD DFT) which has been applied recently to electronic absorption spectra of FBP in many studies [4, 16–18].

Many organic compounds dissolved in glasses exhibit a strong afterglow when excited by ultraviolet (UV) light. This afterglow (phosphorescence) occurs at lower

O. Loboda, *Quantum-Chemical Studies on Porphyrins, Fullerenes and Carbon Nanostructures*, Carbon Nanostructures, DOI: 10.1007/978-3-642-31845-0_1,

frequency than fluorescence (fast emission) and has much longer lifetime (τ_p from 0.1 ms to 100 s) [20]. Though all porphyrins have a high yield of intersystem crossing (a nonradiative transition between low lying singlet and triplet excited states, $S_1 \rightsquigarrow T_1$) only metalloporphyrins exhibit appreciable phosphorescence [8]. Luminescence of porphyrin free bases and their complexes with light metals (Mg, Al) consists of very weak fluorescence; the phosphorescence is of two-three order of magnitude weaker and is very difficult to detect [8, 21–23]. Despite this fact an impressive amount of experimental information has been accumulated on photophysical properties of porphyrin and chlorophyll systems also concerning their triplet state dynamics and spectroscopy [1, 2, 24–30]. Their phosphorescence lifetime and zero-field splitting (ZFS) have been studied by combinations of optical and EPR techniques. The photosynthetic pigments, chlorophyll and bacteriochlorophyll have been studied by the ODMR techniques in vitro and in vivo [27]. The combination of high sensitivity and high spectral resolution makes ODMR methods in zero field well suited to study triplet states in photosynthetic pigments. Fluorescence detection of magnetic resonance (FDMR) is an especially useful technique, since the phosphorescence is often very weak or even undetectable for porphyrins [2]. FDMR was first applied to the bacterial triplet state by Clarke et al. [25]. Following this work a large number of research studies have been reported on the static and dynamic parameters of the triplet state in various species of photosynthetic bacteria [1, 2, 26, 27, 31]. So far, analysis of the complicated ODMR and OD ENDOR spectra of these systems have not been performed at modern ab initio levels of theory. In this chapter we present results from calculations of microwave signals of the ODMR and OD ENDOR spectra in the triplet state of the free base porphin (FBP) molecule by ab initio or DFT approaches.

Although the triplet excited states of photosynthetic pigments are known to be involved in photosynthesis and similar photochemical reaction cycles [27], the calculations have been mostly devoted to the singlet–singlet transitions in the electronic absorption spectra. There are no calculations of the singlet–triplet transitions as far as we know, although triplet state energies have been addressed [16, 18]. In this book we present the first TD DFT attempt to fill this gap by predicting the singlet–triplet transition intensity in the FBP molecule using a recently developed response function technique [32]. As many other porphyrins the FBP molecule does not phosphoresce to any appreciable extent in typical solvents; thus the main progress in experimental studies of their triplet states was achieved with the optical detection of magnetic resonance (ODMR) techniques, especially through the development of nonphosphorescent detected ODMR methods [28].

Our work is also motivated by the long-run ideas that a deeper understanding of fine and hyperfine structure of the triplet state ODMR spectra is necessary for fundamental treatments of photosynthesis and iron porphyrin involvement in oxygen transport and electron transfer in biosystems. For example, spin–orbit coupling (SOC) induced triplet–singlet nonradiative transition (intersystem crossing, ISC) constitutes a very important step in oxygen activation by oxygenases [33–37], and probably in many other enzymatic reactions [38–43]. Electron transfer in photosynthesis and in the bacterial reaction center is accompanied by spin effects (ISC) which could be induced by SOC and by hyperfine interactions [1, 2, 26, 27].

1.1.1 Method of Calculations

Calculations of phosphorescence, zero-field splitting, excitation energies and oscillator strengths for free base-porphyrin were carried out by density functional theory as implemented in the Dalton program [44]. The geometries of the singlet ground state and of the lowest excited triplet states have been optimized for free-base porphin molecule with the D_{2h} symmetry restriction, using the Gaussian program [45] with implementation of the cc-PVTZ basis set [46]. Becke's three-parameter hybrid functional (B3LYP) was used for all calculations except the fine structure calculations. The zero-field splitting parameters were obtained from restricted high-spin open-shell wave functions (ROHF).

1.1.1.1 Calculation of Phosphorescence

The singlet–triplet transition moments and phosphorescence lifetimes were calculated by time-dependent density functions theory utilizing quadratic response (QR) functions; the DFT QR method is described in Ref. [32]. Besides the first excited triplet state (T_1) of the free-base porphin molecule, which has $^3\pi\pi^*$ nature, the second excited (T_2) $^3\pi\pi^*$ state is also studied in order to understand the reasons for correlations of phosphorescence and microwave signals. The rate constant of spontaneous phosphorescence emission $T^a \rightarrow S_0$ from a spin-sublevel (a) is determined by equation:

$$k^a = 2.139\times 10^{10} E^3 \mid M^a \mid^2, \tag{1.1}$$

where E is the $T - S_0$ transition energy, M^a is the electric dipole transition moment (both are in atomic units). The a-axis means the spin quantization axis of the triplet state T^a spin-sublevel in zero external magnetic field, which splittings and axes directions are calculated as the following.

1.1.1.2 Calculation of Zero-Field Splitting

The zero-field splitting in the excited triplet states of polyatomic molecules is described by the tensor D_{ij}, which is widely used to analyze EPR and ODMR spectra. It defines the effective spin Hamiltonian given by the scalar products of the following terms [20]:

$$H_S = \sum_{ij} D_{i,j} S_i S_j, \tag{1.2}$$

where S_i is the ith Cartesian component of the total electron spin operator. In a coordinate system x, y, z with the ZFS eigenfunctions the effective spin Hamiltonian can be written as

$$H_S = -XS_x^2 - YS_y^2 - ZS_z^2, \tag{1.3}$$

where $D_{x,x} = -X$, $D_{y,y} = -Y$ and $D_{z,z} = -Z$. Since the $D_{i,j}$ is a traceless and symmetric tensor ($X + Y + Z = 0$) which is diagonal in its principal axes system it can be described by only two independent parameters, D and E [20]

$$H_S = D\left(S_z^2 - \frac{1}{3}\mathbf{S}^2\right) + E(S_x^2 - S_y^2), \tag{1.4}$$

In Eq. (1.4) the choice of axes is such that the $\mid T_z\rangle$ eigenfunction describes the spin sublevel with the largest splitting; that is, ($\mid Z \mid > \mid X \mid, \mid Y \mid$). In that case the z-axis is the main axis of the ZFS tensor, Eq. (1.4), where

$$D = -\frac{3}{2}Z, \quad E = \frac{1}{2}(Y - X). \tag{1.5}$$

One has to note that the choice of z-axis, which determines the ZFS parameters D and E, depends on the molecule and on the symmetry of the triplet state. In our calculations of the free-base porphin molecule z-axis is along the N–H bonds and x-axis is out of plane. Numeration of atoms and orientation of axes is shown in Fig. 1.1. In that case, as follows from the present and from previous calculations [7, 28], D is not determined by Eq. (1.5), but by $D = -\frac{3}{2}X$, something that coincides with experimental findings [28]. The ZFS tensor can be obtained by contracting two-electron field gradient integrals with a quintet two-electron density [47]

$$D_{ij} = \sum_{tuvw} d_{ij,tuvw} q_{tuvw}, \tag{1.6}$$

where $d_{ij,tuvw}$ is the integral over partially occupied orbitals t, u, v, w. The integral $d_{ij,xxxx}$ is calculated as Cartesian i, j component of the two-electron field gradient operator; the matrix q_{tuvw} is a two-electron density matrix corresponding to the zeroth component of the quintet combination of two triplet operators. We refer to Ref. [47] for other details.

Normally, for the triplet $^3\pi\pi^*$ state of organic molecules the ZFS parameters can be entirely determined as the SSC expectation value [48, 60]. But for the $^3\pi\pi^*$ states in porphin the second-order SOC effect can produce an appreciable contribution to the ZFS since the FBP molecule has lone pairs at nitrogen atoms. We have therefore computed the T_1 state of free-base porphin molecules taking into account both SSC and SOC perturbations. The SSC expectation values are calculated here by the single-determinant SCF method for the high spin open shell [47] using Dalton code.

The spin–orbit contribution to the splitting of the lowest triplet state has been estimated by the second order perturbation theory using six excited states of each symmetry (three triplet and three singlet states) where spin–orbit coupling integrals were calculated by employing the atomic mean-field approximation [49].

1.1.1.3 Calculation of Hyperfine Coupling Constants

For the calculations of the hyperfine coupling (HFC) constants the unrestricted Becke's UB3LYP hybrid functional in combination with the triple-zeta EPR-III basis set [50] implemented in Gaussian 98 program [45] was used. Solvent effects on the hyperfine constants were estimated using the polarizable continuum model (PCM) [51] implemented in Gaussian 98 [45]. The isotropic HFC constant, a, is related to the spin density at the corresponding nucleus by:

$$a = (8\pi/3) g_e g_N \beta_e \beta_N |\Psi_0|^2, \tag{1.7}$$

where g_e is the electron gyromagnetic ratio, β_e is the electron (Bohr) magneton, β_N is the nuclear magneton, g_N is the nuclear gyromagnetic ratio of the corresponding nucleus, and $|\Psi_0|^2$ is the unpaired electron spin density at the nucleus. The spin dipole coupling tensors are characterized by anisotropic interaction and can be described by the hyperfine tensor:

$$T_{ij}(N) = g_e g_N \beta_e \beta_N \sum P_{\mu\nu}^{\alpha-\beta} \left\langle \varphi_\mu | r_{kN}^{-5} (r_{kN}^2 \delta_{ij} - 3 r_{kNi} r_{kNj}) | \varphi_\nu \right\rangle, \tag{1.8}$$

where $P_{\mu\nu}^{\alpha-\beta}$ is an element of the spin density matrix. The Gaussian 98 program provides principal values and directions for the anisotropic contribution. The components of the total hyperfine tensor, **A**, observed in the ESR and ODMR spectra, are obtained as a sum of isotropic HFC constant and anisotropic electron-nuclear spin dipole coupling: $A_{ii} = a + T_{ii}$. Thus, the hyperfine principal values can readily be obtained.

1.1.2 Singlet–Singlet Absorption

The singlet excited states of the FBP molecules calculated with the B3LYP functional and cc-pVTZ basis set by our TD DFT code are in good agreement with the measured electronic absorption spectra of [10, 52]. The first absorption band, namely the Q^0 band in the visible region ($\lambda = 627$ nm), is attributed to the 1^1B_{1u} state in our choice of axis. In respect to the choice of the axes used in previous theoretical works [4, 12, 13, 16, 53] the B_{1u} notation in our work corresponds to the B_{3u} irreducible representation (and vice versa) in many of the other works [4, 13, 16]. Since the direction of the y-axis (N–N direction) coincides with the one in Refs. [4, 13, 16], the B_{2u} irreducible representation is the same. The Q^0 line is polarized along the N–H axis (z direction, Fig. 1.1); its oscillator strength (0.003) is in a reasonable agreement with experiment (0.001) [52]. Sundholm [16] obtains better agreement with the transition energy (2.13–2.24 eV), but worse value for the oscillator strength (0.00018–0.0009). The second Q^0 band ($\lambda = 511.5$ nm) with y-polarization (transition to the 1^1B_{2u} state [4, 16, 54]) is rather well reproduced by our calculation (Table 1.1). The calculated oscillator strength (0.035) is in better agreement with experiment (0.06) [10, 54] than in these other TD DFT treatments (0.00004–0.0006 [16]). The DFT/MRCI calculations by Parusel and Grimme [4] reproduce experimental wavelengths of both the Q^0 bands quite well while for the oscillator strengths their results are less satisfactory.

The vibronic Q^1 bands (575 and 483 nm) are much more intense in the absorption spectrum of the FBP molecule than the 0–0 Q^0 bands (the superscript refers to the upper state vibrational level) [16]. This cannot be attributed to the Franck–Condon factors since one cannot anticipate large displacements in equilibrium geometries between the ground and excited $^1B_{2u}$, $^1B_{1u}$ states as follows from our geometry optimization of the S_0 and T_1 states. Thus the relatively high intensity of the Q^1 bands indicates a strong vibronic perturbation and intensity borrowing from upper transitions.

The very strong band B (Soret band) in the near UV region ($\lambda = 372$ nm) can be attributed to the 2^1B_{1u} and 2^1B_{2u} states in our choice of axis. Two other intense transitions to the 3^1B_{1u} and 3^1B_{2u} states are probably also connected with the Soret band. All these states are approximately in the 3.4–3.6 eV region (3^1B_{1u} is slightly above). Vibronic mixing of these states can lead to some energy shifts and intensity redistribution. There are no other intense transitions in the absorption spectrum of the FBP molecule. In general, our TD DFT linear response calculation of the vertical singlet–singlet electronic absorption spectrum of the FBP molecule is in accord with other recent DFT results [4, 16] and provides better agreement with experiment [10, 54] than SAC-CI [12, 13] or CASPT2 [14] methods. We cannot accept the interpretation of Nakatsuji et al. based on SAC-CI calculations [12, 13] where they assign the 2^1B_{2u} state to the week N-band, which occurs as a shoulder of the Soret band.

1.1.3 Prediction of Phosphorescence Radiative Lifetimes

Solovjev et al. [8, 21] have shown that the main decay route of the first excited singlet state S_1 (1^1B_{1u} in our notations) of free-base porphin is the intersystem crossing to the lowest triplet state T_1 (1^3B_{2u} in our results, Table 1.2) with a quantum yield $\phi_{ISC} = 0.9$. So far no phosphorescence has been observed from free-base porphin in solid solvents usually used for phosphorescence detection; only $T_1 \rightarrow S_0$ emission induced by the external heavy atom (EHA) effect has been recorded [21–23]. Gouterman and Khalil [22] observed a low-resolution phosphorescence spectrum of free-base porphin in ethyl iodide + EPA mixture. The natural radiative lifetime for phosphorescence of various free bases were found to be 70 s and higher. Michl et al. [23] have reported Fourier transform measurements of fluorescence and phosphorescence of FBP in rare gas matrices. They obtained a highly resolved intense phosphorescence spectrum in xenon dominated by the EHA effect.

The calculated energy for the lowest triplet state (1^3B_{2u}), 1.5 eV above the ground state, is in a good agreement with the 0–0 line wavelength (785 nm) in the phosphorescence spectrum [22], which corresponds to 1.58 eV for the T–S transition energy. The triplet excited states calculated by the TD DFT response method are presented in Table 1.3. Two roots of each symmetry are included. The second low-lying triplet state is 1^3B_{1u} which is 0.2 eV higher in energy than the T_1 state. This is in good agreement with other calculations [16, 18].

From an analysis of the microwave-induced fluorescent signal and kinetic equations van Dorp et al. [6] estimated the radiative decay rate of the T^z spin sublevel (in our notation of axes) to be of the order of $2 \times 10^{-3}\,s^{-1}$. This is in excellent agreement with our calculated rate constant for the T^z spin sublevel $k_z = 2.65 \times 10^{-3}\,s^{-1}$, something that illustrates the strength of quadratic response TD DFT theory with spin–orbit coupling and electric dipole moment operators. Calculation of the T–S transition probabilities is a very difficult task [55]; for a very weak spin-forbidden radiation the results strongly depend on the truncation of the summation in the traditional sum-over-sate treatment by perturbation theory. In the quadratic response approach this summation is replaced by solving system of linear equations, and all excited singlet and triplet states relevant to the chosen basis set are taken into account in a size-consistent manner. In the MCSCF QR method the problem of choice of CAS space is difficult as it often significantly influences the calculated phosphorescence rate constant [55]. The QR DFT method seems not to suffer from these difficulties and has provided impressive results so far [32, 56].

The other spin sublevels are non-active in the radiative decay of the T_1 state of free-base porphin in accord with the ODMR measurements [6, 29]. The T^y spin sublevel is completely dark by symmetry selection rules for spin–orbit coupling and electric dipole operators. For the out-of-plane spin sublevel, T^x, the calculated radiative decay rate constant is equal to $0.8 \times 10^{-6}\,s^{-1}$, thus it is practically negligible.

In Table 1.2 the experimental rate constants measured for different spin-sublevels by the fluorescent-detected ODMR techniques [6] are also presented. They are much higher than the calculated radiative decay rate constants. This means that the experimental rate constants, presented in the third row of Table 1.2, refer to the nonradiative decay. A good quantitative agreement between our calculated k^z value and the radiative rate constant of the T^z spin-sublevel, extracted from kinetic analysis of the ODMR signals [6], indicates, without doubt, the correct interpretation of the observed decay. It is also important that an identical interpretation of the most active T^z spin-sublevel is obtained in our theory and in the ODMR experiment [6], where the z-axis coincides with the N–H...H–N bond direction.

The calculated radiative lifetime at the high temperature limit is very long (about 19 min, Table 1.2). At liquid helium temperature, when spin-lattice relaxation is frozen, the phosphorescence radiative lifetime should be 3 times shorter, since only one spin-sublevel is active; it is predicted to be 376 s.

The second triplet state of the $^3B_{2u}$ symmetry is predicted to be more active in SOC-induced mixing between S and T states, which leads to stronger emission from the T_2^z spin-sublevel (Table 1.2). The radiative lifetime of this spin-sublevel is equal to 50 s. Since the energy of the 2^3B_{2u} state is close to the first excited singlet state (Table 1.1) one can anticipate an involvement of the 2^3B_{2u} state in the intersystem crossing $S_1 \rightsquigarrow T_2$, which could be one reason for the very effective intersystem crossing in the FBP molecule.

Spin selective information on the lowest triplet state decay was obtained by optical detection of magnetic resonance transitions between the spin components of the T_1 state of FBP in n-octane. Because of the absence of phosphorescence at these conditions, the ODMR signals were detected via changes in the $S_1 \rightarrow S_0$ fluorescent intensity [6, 29]. Our calculations reproduce the fluorescent frequency and radiative constant rather well. In order to complete the interpretation of the microwave-induced fluorescent ODMR measurements [6, 29] one has to calculate the zero-field splitting in the T_1 state and hyperfine coupling between electron and nuclear spins.

1.1.4 Fine and Hyperfine Structure of the 1^3B_{2u} *State*

In the first microwave-induced fluorescent ODMR experiment van Dorp et al. [24] measured the fine structure of the triplet state of free-base porphin in n-octane Shpolskij-matrix at 4.2 K. They obtained three zero-field signals for one optical site of FBP at 16331 cm^{-1} equal to 390, 1114 and 1494 MHz. These signals correspond to the ZFS parameter values $| D | = 0.0435\,\text{cm}^{-1}$ and $| E | = 0.0063\,\text{cm}^{-1}$, where the main axis of D is proposed to be perpendicular to the plane. From our calculation we get the following parameters: $X = -319.8$ MHz, $Y = 97$ MHz and $Z = 222.7$ MHz, which correspond to $D = -(3/2)X = 0.016\,\text{cm}^{-1}$ and $E = (1/2)(Z - Y) = 0.0021\,\text{cm}^{-1}$. The SOC contribution to ZFS is found negligible. This result, obtained by single-determinantal SCF ROHF calculations for the $^3B_{2u}$ state, gives only qualitative agreement with the experimental ZFS parameters (the corresponding DFT approach is now in progress). Later van Dorp et al. [5] resolved hyperfine coupling (HFC) structure by the EPR experiment in the same host single crystal at 1.3 K. They have obtained a rather large isotropic HFC constant (3.92 Gauss) for the bridging methine protons [5] (numbers 25–28 in our Fig. 1.1). From these results they concluded that the HOMO is of $5b_{1u}$ symmetry; by comparison with the calculated ZFS parameters the authors of Ref. [5] concluded that the upper orbital is $4b_{3g}$ in agreement with later DFT results [16, 18]. The calculated HFC constants are presented in Table 1.4 together with experimental data. When our calculations were completed the recent, important, work of Kay [57] come to our attention in which the HFC structure from the time-resolved electron-nuclear double resonance (TR ENDOR) spectrum of FBP [57] was determined. He found HFC constants for all protons besides the methine protons and calculated also the protonic HFC structure by unrestricted B3LYP DFT with the EPR-II basis set and assuming a triplet state geometry optimized with the 6-31G** basis [57]. His results for isotropic HFC constants are quite close to ours (Table 1.4) though the anisotropic dipole contributions are rather different. For proton number 3 in his notation (numbers 25–28 in our Fig. 1.1) Kay calculated the total HFC tensor component in the perpen-

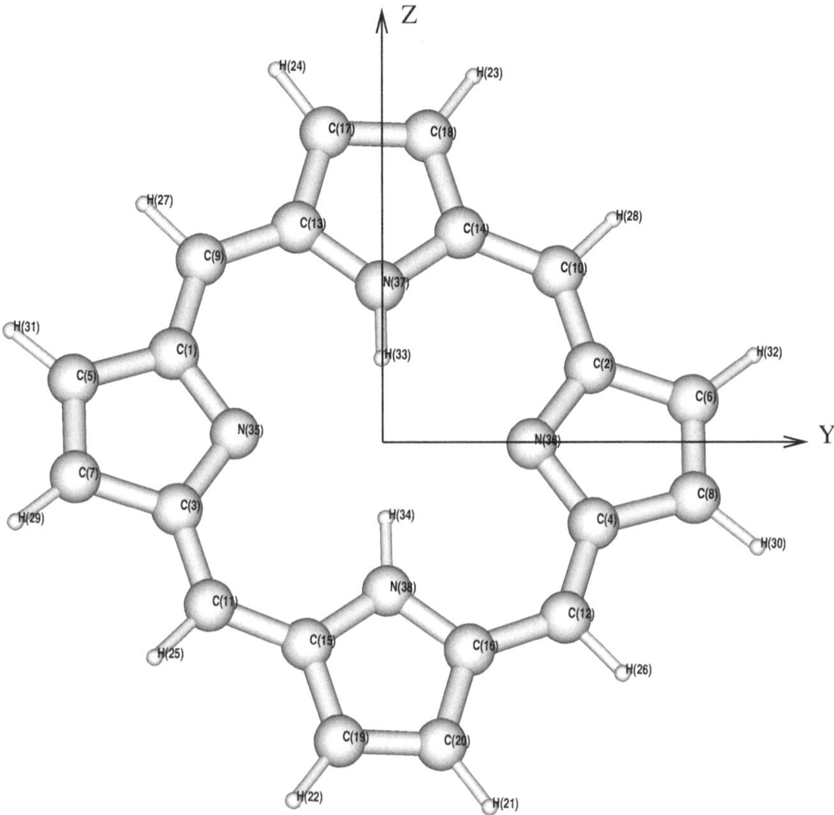

Fig. 1.1 Numeration of atoms in the FBP molecule and choice of axes

dicular direction equal to -0.01 MHz because of cancellation of isotropic (0.72 MHz) and anisotropic (-0.73 MHz) contributions [57]. In our larger basis set we received a total A_{xx} value (-0.90 MHz) in good agreement with the ENDOR measurements (-0.72 MHz). For the bridging methine protons our calculation provides an isotropic HFC constant (-12.53 MHz) in agreement with Kay's calculation (-13.35 MHz) which can be compared with the value determined by van Dorp et al. [5] from EPR measurement to be 11 ± 0.6 MHz. The theoretical agreement for two independently optimized geometries of the triplet state with different basis sets establishes beyond doubt the negative sign of this HFC constant.

Information about the hyperfine coupling on the ^{13}C nuclei is known from experiments. Our calculation predicts a very high positive anisotropic dipole contribution for the perpendicular component, $A_{xx} = 74$ MHz at the bridging methine carbons C(9)–C(12). This is again in agreement with the negative sign of the hyperfine coupling constant for neighboring bridging methine protons.

The analysis of van Dorp et al. [5] of the EPR spectrum indicated that the hyperfine coupling to the ^{14}N nuclear spins proves to be very small. This is supported by our calculation for N(37), N(38) atoms which are connected with protons. All known geometry optimizations of the lowest triplet state in the FBP molecule [16, 18, 57] provide a D_{2h} structure in agreement with our result. We also have performed geometry optimization for the lowest triplet state of the *cis*-tautomer, when two N–H groups are sitting on neighboring pyrrole rings connected by the C(9) methine bridge and hydrogen atoms are bonded with N(37) and N(35) atoms (Fig. 1.1). The triplet state energy of the *cis*-tautomer is 12.5 kcal mol^{-1} higher than the "normal" triplet state with the *trans*-structure (Fig. 1.1) and is lower by 4.6 kcal mol^{-1} than the "normal" S_1 state. A photoinduced shift of two inner protons in free-base porphin was discovered by Solovjev et al. [21] 30 years ago, but the mechanism of the process is still unknown [9, 23]. It was proposed that the photoisomerisation reaction may be associated with vibrational relaxation that follows the intersystem crossing from the S_1 state to the triplet manifold [9, 23, 29]. Our results for the *cis*-tautomer support the qualitative energy diagram proposed in Ref. [9].

The C=C and C–C bonds redistribution in methyne and pyrrole moieties constitutes the main structural changes upon the $S_0 - T_1$ transition. In the ground S_0 state the C(14)–C(10) bond length, 1.388 Å, is almost equal to the C(10)–C(2) bond length, 1.395 Å, while in the T_1 state they are rather different, 1.423 and 1.383 Å, respectively, see Table 1.6. In the pyrrole ring (with N–H group) the C(17)–C(18) and C(18)–C(14) bond lengths are very different in the ground state, 1.366 and 1.430 Å, respectively, but in the triplet state they are very close, 1.397 and 1.402 Å, respectively. The C–N distance is almost unchanged. In the other rings (without N–H group) there are no large changes upon the $S_0 - T_1$ transition (Table 1.6). In this respect the structural question arises: why the ENDOR signals from the α protons H(29)–H(32) on the azomethine rings (without N–H group) which are denoted by number 3 in Kay's notation and on the α' protons H(21)–H(24) of the pyrrole ring, which are denoted by number 2 in Kay's notation, are so different? We find that this can be explained by the changes in geometrical parameters of the C–C skeleton upon $S_0 - T_1$ $(\pi - \pi^*)$ excitation, and also by the large differences in the hyperfine coupling on the ^{13}C nuclei of the α and α' types. In our choice of axes the lowest triplet state is connected with the $5b_{3u} \rightarrow 4b_{1g}$ excitation, which corresponds to the $5b_{1u} \rightarrow 4b_{3g}$ transition in the notation of Refs. [5, 16, 58]. At this excitation a large π-spin-density (0.407) is created at the methine carbons C(9)–C(12) with simultaneous bond order redistribution which also follows from the simple MO-picture [5, 10]. In terms of McConnell relation [59] this explains a large negative value of isotropic HFC constant at the methine protons H(25)–H(28). However, a simple qualitative MO-picture, see e.g. Ref. [5, 10], cannot explain a quantitative difference in HFC constants for protons of groups number 2 and 3. Geometry optimization of the lowest triplet state and account of polarization functions in the cc-pVTZ basis set are important in this respect.

Another intriguing question has been raised by Kay: why the ENDOR signals from the α protons, H(29)–H(32), and on the N–H protons have small line widths, while the signal from α' protons, H(21)–H(24), are broadened into a doublet [57]?

This nonequivalence indicates a symmetry reduction to the C_{2h} point group in the solid environment [57]. Kay tried to reproduce it by adding nonsymmetric point charges near protons H(22), H(23), H(30) and H(31) placed 3 Å above the porphin ring in his DFT simulation. He received almost the same shift of isotropic HFC constants for protons of groups number 1 and 2, the former being slightly lower. For protons of group number 3 the shift was negligible, as it was not seen in the ENDOR spectrum, which partly supported the proposed explanation [57].

1.1.5 Solvent Effects on the Hyperfine Coupling

We try to account for the solvent effect on the hyperfine coupling by using the polarizable continuum model (PCM). A large reduction of the isotropic hyperfine coupling on the ^{13}C and ^{14}N nuclei is obtained in toluene and in water solvents (Table 1.5) in comparison with the gas phase calculation (Table 1.4). At the same time there are no large changes for the HFC tensors on protons. For the bridging methine protons the PCM calculation in toluene predicts a better agreement for the isotropic HFC constant (−11.3 MHz) with the experimental value obtained in n-octane (if negative sign is accepted) [5]. We note here that n-octane and toluene have very similar dielectric constants. For nitrogen atoms N(35, 36) which have lone pairs one could anticipate some notable effect for the HFC tensor on going from polar to nonpolar solvents. In fact, neither isotropic nor anisotropic HFC constant on nitrogen atoms (Table 1.5) are changed comparing the toluene and water solvents.

The large difference in the A_{xx} constants for the protons of two groups (number 2 and 3) is still reproduced for all solvents (Tables 1.4, 1.5). Since the PCM model does not influence the symmetry of the dissolved molecule we cannot reproduce the splitting of signals on the nuclei H(21)–H(24). However, at least we can see that the solvent effect does not lead to appreciable broadening of the ENDOR signal on these protons. This is in agreement with the model of nonsymmetric intermolecular perturbations proposed by Kay [57].

1.1.6 Discussion

An aspect of this work in a longer perspective is connected with theoretical analysis of the external magnetic field effects on enzymatic reactions and photosynthesis. The photoinduced triplet EPR signal in the bacterial reaction center is strongly polarized [1, 2]. While the spin polarization of the triplet state in high magnetic fields is a general phenomenon [27, 60], the spin polarization of the bacterial triplet state is very unusual; it can not be derived from a normal inter-system crossing in a single molecule [2]. Several mechanisms have been proposed [27, 61], but the problem cannot be solved without complete information about reliable wave functions with account of magnetic perturbations at an ab initio level.

The traditional EPR technique uses the fixed microwave frequency at the cavity resonance and an external magnetic field (**B**) being swept. Since biopolymers are not usually available as oriented samples there are many disadvantages of such EPR applications. The triplet spin sublevel properties are determined by intramolecular spin–orbit coupling (SOC). Thus it is the ZFS sublevels that acquire the divergent properties (such as differences in lifetimes) which makes the application of ODMR methods possible. In a randomly oriented sample, the applied magnetic field **B** mixes the ZFS states and results in a weaker optical signal. Moreover, in a strong magnetic field **B** the spin-lattice relaxation rate increases and reduces the spin alignment of the triplet state [62]. Thus the most intense ODMR signals will be observed in zero field. The ZFS lines are only about 1 MHz in width in ODMR spectra of molecular solids and they are independent of whether the sample is polycrystalline or oriented.

All ODMR studies of biopolymer triplet states have been made in zero field using polar glass solvents [27]. The ZF ODMR lines of biopolymers are much broader (100 MHz) than those observed in molecular crystals. Broadening of the ODMR lines as well as of the phosphorescence lines arise from heterogeneity in the local environment of the emitting chromophores [63, 64]. The aim of the present project was to establish the nature of the ODMR line broadening for triplet state signals of porphyrins and their derivatives. We have made a first attempt to establish direct connections between the optical and magnetic properties of the triplet states on the ground of detailed knowledge about the electronic structure of their emitting chromophores. Calculations of radiative lifetime of the triplet state spin sublevels with account of intramolecular and intermolecular spin–orbit coupling effects by quadratic response DFT methods with simultaneous estimations of the ZFS frequency and other parameters of the spin-Hamiltonian (isotropic and anisotropic hyperfine coupling tensors, g-factors, nuclear quadrupole resonance parameters) provides useful information, which is not available from experimental ODMR and EPR studies of biopolymer triplet states. This can hopefully complement experimental data analysis and increase the power of the ODMR techniques in zero field.

The high external magnetic field can mix the zero-field spin sublevels and therefore will change the rate constants of the T–S transitions [65]. This concept can be applied to triplet chromophores in biopolymers with slow rotational diffusion. Moreover, the Zeemann perturbation can induce additional mechanisms for the T–S mixing [65, 66]. The standard Liouville equation for spin density matrix [65] can be used for prediction of magnetic field effects on spin-selective photoprocesses and enzymatic reactions if the ZFS and hyperfine coupling parameters are properly interpreted in connection with the T–S transitions rate constants.

1.1.7 Conclusions

All parameters of the effective spin Hamiltonian measured in magnetic resonance experiments—ODMR, EPR and ENDOR—on the lowest triplet state of the free-base porphyrin molecule, have in the present work been interpreted together with spin-selective radiative rate constants. The recently derived quadratic response time-

Table 1.1 Singlet excitation energies, E (eV), oscillator strengths (in parentheses), and transition dipole moments, M_a (a.u.), of the FBP molecule

State	This work		SAC-CI[a]	TD DFT[b]	DFT/MRCI[c]	Experiment[d]
	E(f)	M_a	E(f)	E(f)	E(f)	E(f)
1^1B_{1u}	2.25(0.003)	$-0.2378(M_z)$	1.7(0.0002)	2.24(0.00018)	1.94(0.0007)	1.98(0.01)
1^1B_{2u}	2.35(0.035)	$0.7830(M_y)$	2.19(0.0006)	2.39(0.00004)	2.38(0.0014)	2.42(0.06)
2^1B_{1u}	3.32(0.641)	$2.805(M_z)$	3.43(1.1)	3.27(0.40)	3.07(0.4846)	3.33(1.15)
2^1B_{2u}	3.39(0.2238)	$1.639(M_y)$	3.62(1.87)	3.45(0.61)	3.17(0.6604)	
3^1B_{2u}	3.62(0.807)	$3.014(M_y)$	4.36(0.437)	3.70(0.55)	3.54(0.862)	3.65(<0.1)
3^1B_{1u}	3.96(0.545)	$-2.369(M_z)$	4.08(1.09)	3.79(0.82)	3.56(0.9826)	
1^1B_{3u}	3.82(0.0007)	$0.0860(M_x)$	4.51(0.0005)	4.03(0.001)		
2^1B_{3u}	5.87(0.0023)	$0.1281(M_x)$				
3^1B_{3u}	6.64(0.0016)	$-0.1008(M_x)$				

[a]Reference [12], [b]Ref. [16], [c]Ref. [4], [d]Ref. [52]

Table 1.2 Excitation energies (eV), transition moments (a.u.), rate constants k^a (s^{-1}) for emission from T^a spin sublevel and phosphorescence lifetime τ (s) for the low lying triplet states of $^3B_{2u}$ symmetry

State	E	$\langle ^3B^x_{2u}\|z\|^1A_g\rangle$	$\langle ^3B^z_{2u}\|x\|^1A_g\rangle$	k^x	k^y	k^z	τ_{av}[c]
1^3B_{2u}	1.51	0.4×10^{-7}	0.27×10^{-4}	0.76×10^{-6}	0	0.26×10^{-2}	1128.42
1^3B_{2u}Exp[a]	1.58					$\approx0.20\times10^{-2}$	
1^3B_{2u}Exp[b]				6.0	75	230	
2^3B_{2u}	2.01	0.1×10^{-5}	0.5×10^{-4}	0.90×10^{-5}	0	0.02	140.47

[a]Reference [6]: the radiative k^z rate constant determined in a non-direct way, [b]Ref. [27]: the observed (nonradiative) rate constants, [c]τ_{av}—averaged phosphorescent radiative lifetime in high-temperature limit

dependent density functional method [32] has been employed for calculations of phosphorescence radiative lifetimes. For the most active T^z spin sublevel, where the z-axis coincides with the N–H...H–N bond direction, the radiative lifetime is obtained as high as 377 s in good agreement with an indirect experimental estimation (an order of 500 s) obtained from kinetic analysis of the fluorescence-detected ODMR signals [6]. The other spin sublevels are practically dark in complete agreement with results from the kinetic equations of van Dorp et al. [6] used for analysis of his ODMR experiment. Such a slow radiative decay is very unusual for a molecule which have nitrogen lone pairs and a number of $n\pi^*$ states in the near UV region. The correct prediction of the spin quantization axis of the most active spin sublevel and of its radiative lifetime in the lowest triplet state of the FBP molecule must be taken as a proof of capability of the quadratic response time-dependent DFT theory; it explains the old puzzle why the FBP phosphorescence has never been observed in the absence of the external heavy atom effect.

We have also interpreted the zero-field splitting (ZFS) and hyperfine coupling (HFC) parameters of the triplet state. The lowest spin sublevel T^x has the spin quantization axis perpendicular to molecular plane in agreement with ODMR and EPR experiments. Our interpretation of the HFC constants for protons accords in the

Table 1.3 Triplet excitation energies (eV)

State	This work	Exp[a]	Nguyen[b]	Sundholm[c]
1^3B_{2u}	1.51	1.58	1.53	1.58
1^3B_{1u}	1.70		1.78	1.74
2^3B_{2u}	2.01			1.97
2^3B_{1u}	2.15			1.99
1^3B_{3g}	2.68		3.34	2.56
1^3A_g	3.25	3.17[d]	3.18	2.63
2^3B_{3g}	3.40	3.23[e]	3.66	2.97
2^3A_g	3.44		3.82	3.28
1^3B_{2g}	3.74			3.19
1^3A_u	3.77			3.19
1^3B_{1g}	4.06		3.96	3.14
1^3B_{3u}	4.08			3.14

[a]References [21–23], [b]Ref. [18], [c]Ref. [16]: the B_1 and B_3 notations are interchanged corresponding to our axis, [d]the first observed T–T band (800 nm) corresponds to the $1^3B_{2u} \rightarrow 1^3A_g$ transition in agreement with experiment [21–23] and with the assignment of Nguyen et al. [18], [e]we assign the second T–T absorption band (752 nm) to the $1^3B_{2u} \rightarrow 2^3B_{3g}$ transition, while Nguyen et al. [18] have made their assignment to the first 1^3B_{3g} state. At the same time our result for the 1^3B_{3g} states is in agreement with the data of Sundholm [16]

main features with the recent results of Kay [57]. A very weak hyperfine coupling at the ^{14}N nuclei connected with protons is proved in accordance with the observation of van Dorp et al. [5] Quite strong solvent effects on the isotropic hyperfine coupling constants for the ^{13}C and ^{14}N nuclei are predicted on the ground of the polarizable continuum model. A conclusion of the present work is that the TD DFT response method is proved to be very useful for studying the spin-selective properties of the lowest triplet state of free-base porphin.

1.2 Isomerisation of Free-Base Porphin

The fluorescence spectrum of H_2P in an n-octane single crystal shows a doublet structure which arises from a superposition of two almost identical spectra displaced by $65\,cm^{-1}$ [6, 24]. These two spectra result from the presence of two distinct H_2P molecules in n-octane. It was established that there exist two sets of the H_2P molecules which occupy equivalent positions in the host crystal related to different orientations of the N–H...H–N axis [9]. While the porphin planes are parallel for the two sites, the N–H...H–N axes are at right angles. Solov'ev [67] succeeded to observe changing of the location of central protons from one set of opposite nitrogens to the other when the tetrabenzoporphyrin (TBP) molecules were excited by light at the temperature 77 K. Meanwhile in the dark the transformations of central protons have not been registered so far. This fact assumes that isomerisation of H_2P molecules takes place in the excited S_1 or in the lowest T_1 state.

Table 1.4 Hyperfine coupling constants (MHz) for the $^3B_{2u}$ state of FBP molecule

HFC	^{1}H						^{13}C				^{14}N
N^a	21–24	25–28	29–32	33–34	1–4	5–8	9–12	13–16	17–20	35–36	37–38
N^b	2	1	3	4	α	β	m	α'	β'		
a_N^a	−3.65	−12.53	0.66	−1.37	−12.88	−0.017	13.58	−5.95	1.58	5.81	0.11
a_N DFT[b]	−3.88	−13.35	0.72	−1.83							
a_N Exp[c]		11 ± 0.6									
$(T_{xx})_N^a$	−1.74	−0.77	−1.56	−5.02	−15.13	−3.62	60.85	6.17	16.36	26.58	3.12
$(T_{xx})_N$ DFT[b]	−1.05	−0.88	−0.73	−2.71							
$(A_{xx})_N^a$	−5.39	−13.30	−0.90	−6.39	−28.01	−3.63	74.43	0.22	17.94	32.39	3.23
$(A_{xx})_N$ DFT[b]	−4.69	−14.23	−0.01	−4.54							
$(A_{xx})_N$ Exp[c]	−3.93/ − 4.33		−0.72	−3.35							

[a]This work, [b]Ref. [57], [c]Ref. [5]

Table 1.5 Hyperfine coupling constants (MHz) calculated for the $^3B_{2u}$ state of FBP molecule in solvents

Solvent	HFC	^{1}H						^{13}C				^{14}N
	N^a	21–24	25–28	29–32	33–34	1–4	5–8	9–12	13–16	17–20	35–36	37–38
		2	1	3	4		α	β	m	α'	β'	
Toluene	a_N	−3.51	−11.31	0.45	−1.41	−4.31	0.05	4.10	−2.25	0.42	2.16	0.03
	$(T_{xx})_N$	−1.85	−1.09	−1.56	−5.08	−14.40	−3.53	62.74	6.72	16.81	27.00	3.11
	$(A_{xx})_N$	−5.36	−12.40	−1.11	−6.49	18.71	−3.48	66.84	4.47	17.23	29.16	3.14
Water	a_N	3.50	−11.74	0.41	−1.38	−4.22	0.004	4.04	−2.22	0.40	2.10	0.02
	$(T_{xx})_N$	−1.86	−1.15	−1.57	−5.03	−13.76	−3.28	62.68	6.88	16.72	26.12	3.01
	$(A_{xx})_N$	−5.36	−12.89	−1.16	−6.41	−17.98	−3.24	66.72	−4.86	17.12	28.22	3.03

[a] The first row corresponds to numeration of atoms in Fig. 1.1. The second row corresponds to numeration of atoms in Ref. [57]

Table 1.6 DFT calculated bond lengths (Å) and bond angles (°) for the ground and triplet excited states of FBP molecule

	^1A_g	$^1A_g^a$	$^3B_{2u}$	$^3B_{2u}^a$
C_{17}–C_{18}	1.366	1.372	1.397	1.401
C_{18}–C_{14}	1.430	1.435	1.402	1.408
C_{14}–C_{10}	1.388	1.394	1.423	1.428
C_{10}–C_2	1.395	1.400	1.383	1.389
C_2–C_6	1.456	1.460	1.459	1.463
C_6–C_8	1.350	1.356	1.347	1.353
C_{14}–N_{37}	1.367	1.372	1.367	1.372
C_2–N_{36}	1.358	1.364	1.367	1.372
C_{18}–H_{23}	1.077	1.082	1.076	1.082
C_{10}–H_{28}	1.081	1.086	1.082	1.087
C_6–H_{32}	1.078	1.083	1.078	1.083
N_{37}–H_{33}	1.010	1.015	1.009	1.015
$\alpha(C_{17}$–C_{18}–$C_{14})$	108.0	108.0	107.5	107.6
$\alpha(C_{18}$–C_{14}–$N_{37})$	106.5	106.5	107.2	107.3
$\alpha(C_{14}$–N_{37}–$C_{13})$	110.9	110.9	110.3	110.3
$\alpha(H_{33}$–N_{37}–$C_{14})$	124.5	124.6	124.8	124.9
$\alpha(N_{37}$–C_{14}–$C_{10})$	125.6	125.5	124.6	124.6
$\alpha(C_{14}$–C_{10}–$C_2)$	127.1	127.1	128.1	128.1
$\alpha(C_{14}$–C_{10}–$H_{28})$	115.9	115.9	115.1	115.0
$\alpha(C_{10}$–C_2–$N_{36})$	125.6	125.5	125.3	125.1
$\alpha(N_{36}$–C_2–$C_6)$	110.8	111.1	110.5	110.8
$\alpha(C_4$–N_{36}–$C_2)$	105.8	105.4	105.7	105.4
$\alpha(C_2$–C_6–$C_8)$	106.2	106.2	106.6	106.5
$\alpha(C_2$–C_6–$H_{32})$	125.4	125.4	125.2	125.2

[a]Reference [18]

Photoinduced double proton transfer in porphin was discovered many years ago [9, 67]. It is also well known that many compounds apart from porphin exhibit this type of reaction (alpha naphthol, methyl salicylate). Volker and van der Waals [9] proposed that the reaction of photoisomerisation associated with the intersystem crossing from the S_1 state into the triplet manifold leads to a shift of the inner protons due to vibronic coupling between the two lower triplet states. Indeed from our previous time dependent DFT calculations [68] it was shown that the first excited singlet and triplet states in H_2P molecule are separated only by 0.75 eV, whereas the two lowest triplet states lie even closer and the energy gap between them is just about 0.2 eV. However, until now the mechanism of the excited-state proton transfer in free base porphin is not clear. In particular, it is not clearly understood whether this occurs in a concerted way by a simultaneous two-proton shift or via a less stable *cis*-tautomer, where two N–H groups are sitting on neighboring pyrrole rings (see Fig. 1.2).

In the present work, we performed DFT calculations to investigate these two possible pathways of isomerisation.

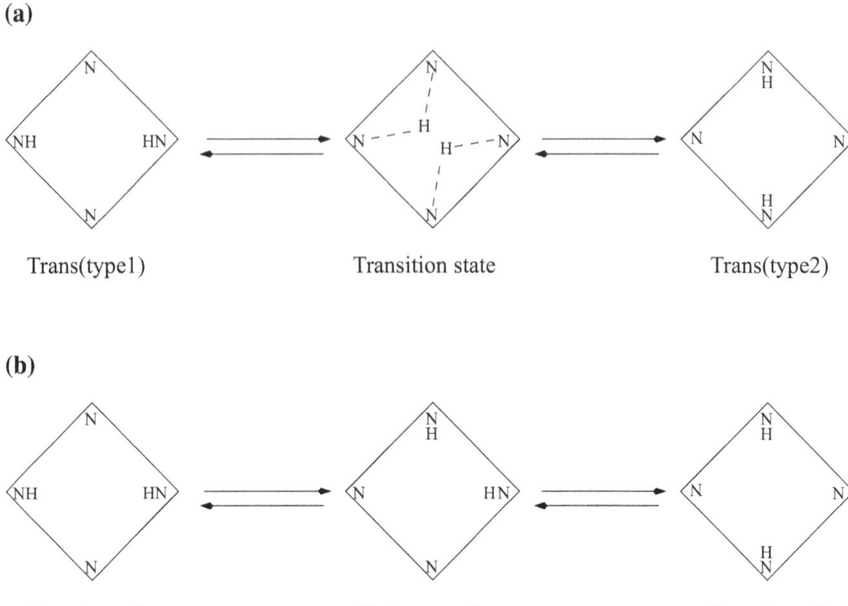

Fig. 1.2 The two mechanisms of isomerisation considered: **a** concerted, **b** stepwise

1.2.1 Method of Calculations

All calculations were performed using Becke's three parameter hybrid density functional method [69] (B3LYP) as implemented in the Gaussian98 program package [45].

Geometries were optimized using the 6-31G(d, p) basis set. Based on these geometries, single point calculations were done using the larger 6-311+G(2d, 2p) basis set in order to obtain more accurate energies. It turned out that the difference between these two basis sets is less than 1 kcal mol^{-1} for relative energies between ground state and transition state (TS).

The triplet state geometry of the *trans* conformation of H_2P was optimized with the D_{2h} symmetry restriction; the *cis* conformation was optimized in C_{2v} symmetry, meanwhile transition state calculations were done without symmetry elements. Potential energy curves for transformation from the *trans* form into the *cis* conformation via transition state were obtained by freezing inner N–H bond and optimizing the rest of the molecule. Zero-point corrections were evaluated at the B3LYP 6-31G(d, p) level of theory.

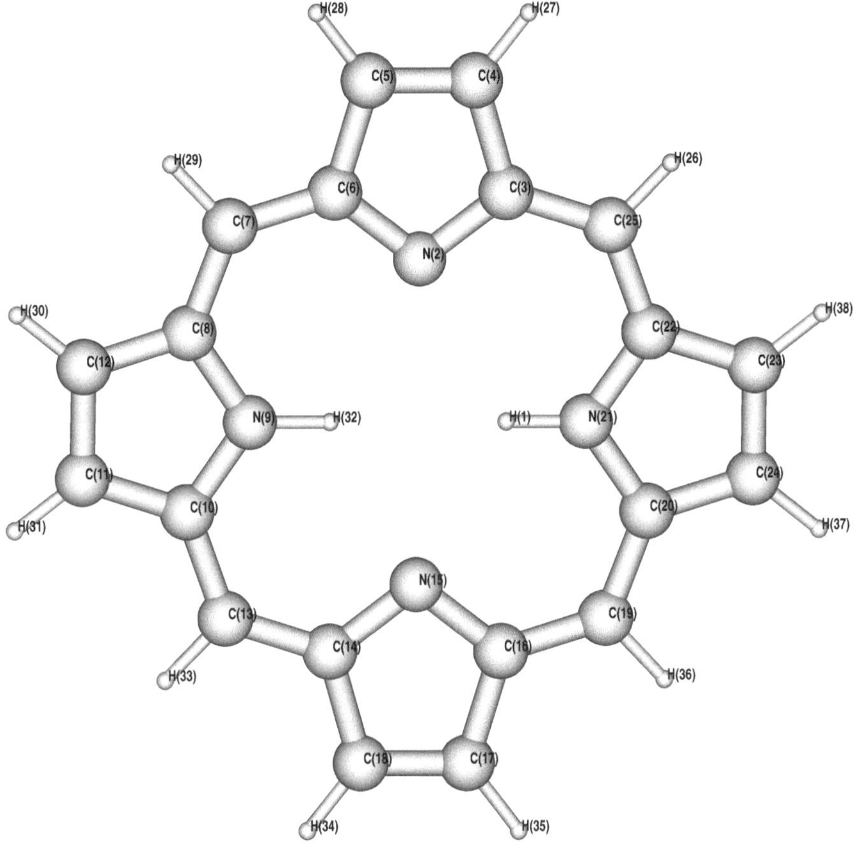

Fig. 1.3 The B3LYP/6-31G(d, p) optimized geometry of the *trans* isomer on the T_1 surface

1.2.2 Results and Discussion

The proton shift of the free base porphins has been studied by ^{1}H and ^{13}C N.M.R. in liquid solution [70, 71].In its stable *trans* conformation, the two protons of H_2P are attached to opposite nitrogens (Fig. 1.3).

During photoreaction they may switch from one pair of nitrogens to the other. Table 1.7 lists the B3LYP/6-31G(d, p) geometrical parameters for optimized stationary points (*cis*, *trans*, TS) of H_2P for both ground singlet and excited triplet states.

The computed bond lengths of bridged carbons in triplet state is 1.43 Å meanwhile the same bond lengths for the singlet ground state is 1.40 Å. The same trends one can observe also for the *cis* and TS structures (Figs. 1.4, 1.5).

Going from *trans* to *cis* conformation within one chosen multiplicity it is easy to observe tendency of porphyrin ring distortion. Thus the C(20)–C(19)–C(16) angle in the *trans* isomer is 128° whereas in the *cis* isomer this angle has the value of 133°,

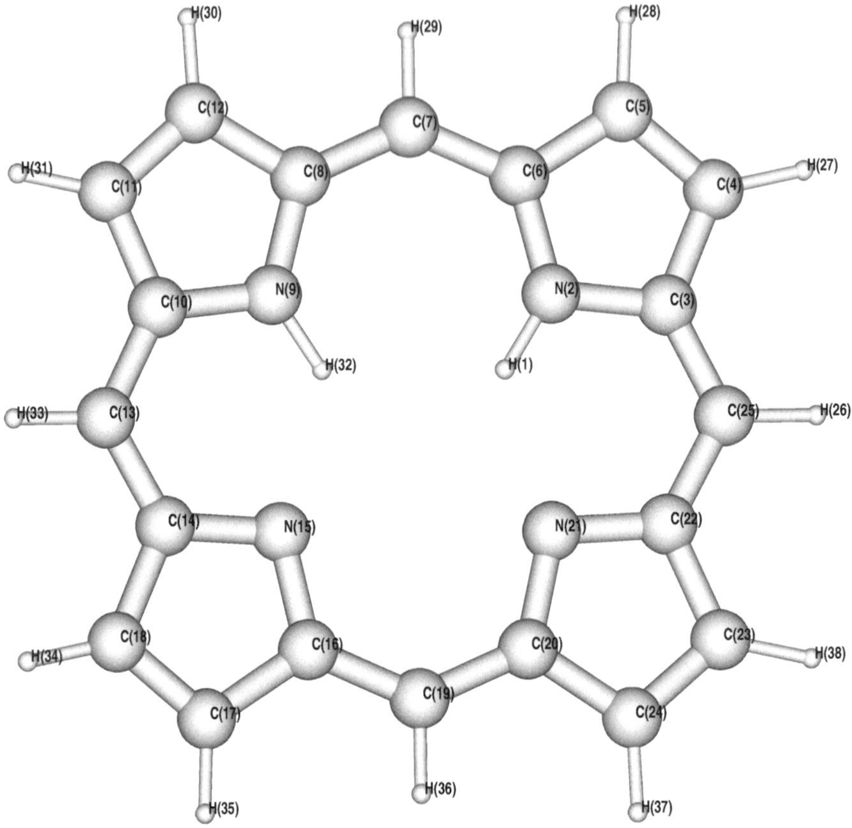

Fig. 1.4 The B3LYP/6-31G(d, p) optimized geometry of the *cis* isomer on the T_1 surface

meanwhile another C(22)–C(25)–C(3) angle decrease from 128° (in *trans*) to 124° (in *cis*). The spin density of carbon (C_7) connecting pyrrole rings decreases dramatically from $\rho = 0.40$ in *trans* to $\rho = 0.28$ and $\rho = 0.17$ in *TS* and *cis*, respectively. Such kind of behavior is very typical for a proton transfer. This transformation may occur in a stepwise fashion through the "cis" isomer or by simultaneous shifts of two central protons.

In the present study we have investigated both these probabilities for both the ground state singlet (S_0) and excited triplet state (T_1). Starting from optimized triplet structure of the *trans* isomer we simulated the reaction pathway by elongating one of the N–H bonds with the initial step 0.1 Å. During the *trans* $\rightarrow$ *cis* rearrangement, the N–H bond cleaves and a new bond is forming with the next adjacent nitrogen. The *trans* $\rightarrow$ *cis* interconversion in the triplet state requires overcoming the barrier of 17.6 kcal mol^{-1}, while the activation energy of the reverse *cis* $\rightarrow$ *trans* teleomerization is only about 5.1 kcal mol^{-1}. We refined our results with the extended 6-311+G(2d, 2p) basis set. It was found that the calculated energies, geometrical

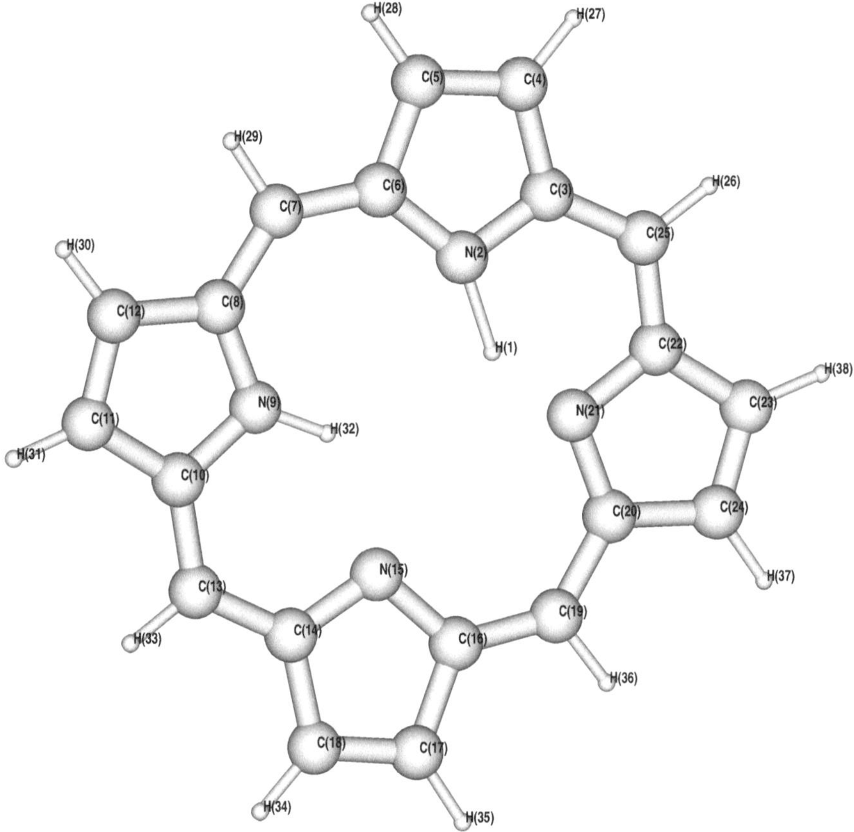

Fig. 1.5 Optimized transition state structure for the isomerisation reaction

structures and vibrational frequencies of all species have negligible discrepancies using the two kinds of basis sets (see Table 1.8). For instance,the deviation in energy is no more than 0.8 kcal mol^{-1}. The relative energies for optimized structures of H_2P calculated at the B3LYP/6-31G(d, p) level are listed in Table 1.8.

Frequency calculations for the transition state (TS) gives a single imaginary frequency ($-1514\,cm^{-1}$) which indicates a large amplitude in-plane motion of the inner proton. It was observed from the experiment [23] that as the temperature increases, transitions between different minima become possible due to the interactions with the thermal bath and the reaction yield increased considerably. At low temperatures the teleomerization is also possible. However a direct transition from *trans* to *cis* form is complicated because of high energy barrier. We assume that at low temperature conditions, the molecule is able to reach the *cis* minima of the triplet state via intersystem crossing since it is higher than the normal *trans* minima. We also have optimized the same isomer structures of H_2P in the ground singlet state. Surprisingly, the energy barrier was found to be 2.5 kcal mol^{-1} lower than the same barrier

Table 1.7 DFT calculated bond lengths Å and bond angles (°) for the ground and triplet excited state of free-base porphin molecule

	Ground state			Triplet state		
	trans	*cis*	*TS*	*trans*	*cis*	*TS*
C_{23}–C_{24}	1.37	1.36	1.36	1.40	1.37	1.38
C_{24}–C_{20}	1.43	1.47	1.45	1.40	1.44	1.43
C_{20}–C_{19}	1.39	1.40	1.40	1.43	1.40	1.44
C_{19}–C_{16}	1.40	1.40	1.40	1.39	1.40	1.38
C_{16}–C_{17}	1.46	1.46	1.46	1.46	1.44	1.46
C_{17}–C_{18}	1.36	1.36	1.36	1.35	1.37	1.36
C_{20}–N_{21}	1.37	1.36	1.37	1.37	1.37	1.37
C_{16}–N_{15}	1.36	1.36	1.35	1.37	1.37	1.38
C_{24}–H_{37}	1.08	1.08	1.08	1.08	1.08	1.08
C_{19}–H_{36}	1.09	1.09	1.09	1.09	1.09	1.09
C_{17}–H_{35}	1.08	1.08	1.08	1.08	1.08	1.08
N_{21}–H_1	1.01	1.92	1.35	1.01	1.91	1.35
$\alpha(C_{23}–C_{24}–C_{20})$	108.00	106.60	107.80	107.56	106.85	107.60
$\alpha(C_{24}–C_{20}–N_{21})$	106.58	110.44	108.84	107.31	110.30	109.61
$\alpha(C_{20}–N_{21}–C_{22})$	110.81	106.03	106.75	110.25	105.73	106.27
$\alpha(N_{21}–C_{20}–C_{19})$	125.50	127.08	130.75	124.49	126.23	129.82
$\alpha(C_{20}–C_{19}–H_{36})$	115.92	114.74	113.59	115.08	113.58	112.34
$\alpha(C_{20}–C_{19}–C_{16})$	127.09	130.53	131.72	128.08	132.84	133.21
$\alpha(C_{19}–C_{16}–N_{15})$	125.53	127.08	126.07	125.23	126.23	125.65
$\alpha(C_{16}–N_{15}–C_{14})$	105.43	106.03	106.15	105.35	105.73	106.03
$\alpha(N_{15}–C_{16}–C_{17})$	111.12	110.44	110.63	110.85	110.30	110.01
$\alpha(C_{16}–C_{17}–C_{18})$	106.17	106.60	106.41	106.48	106.85	106.79
$\alpha(C_{16}–C_{17}–H_{35})$	125.37	125.24	125.37	125.21	125.29	125.16
$\alpha(H_1–N_{21}–C_{20})$	124.59	–	–	124.87	–	147.98

Table 1.8 Relative energies of H_2P isomers and transition state (TS) in kcal mol^{-1}

Conformation	Triplet state		Ground state
	6-31G(d,p)	6-311+G(2d,2p)	6-31G(d,p)
Trans	0.0	0.0	0.0
TS	17.6	18.3	15.2
Cis	12.6	12.4	8.2

for the first triplet state. Existence of a *cis* tautomer in the ground state was proved by the N.M.R. technique [72]. Detection of this isomer in the excited triplet state is a rather hard task since such species is expected to have a quite short lifetime and might be obscured in absorption by other higher populations of the triplet state of the *trans* isomer. At the same time the phosphorescence of the *cis* form might be very small. From the N.M.R. experiment [71] it is known that the barrier in the ground state potential lies 10 kcal mol^{-1} above the energy of the *trans* conformation. This is in a qualitative agreement with our calculations. It is known from Fourier transform fluorescence and phosphorescence of porphin in Ar, Xe matrices [23] that the rate of the dark ground-state reaction is totally negligible. The question why the photoisomerisation reaction does not take place on a potential energy surface of the ground singlet state, which is energetically more favorable in comparison with the excited triplet state, is still open. This is currently under investigation by us. We studied also the alternative pathway of photoisomerisation when two protons move simultaneously in a concerted manner. In this procedure two protons were fixed symmetrically between two neighbor N atoms while the remaining atoms were optimized. Optimization of such a transition state always converged to the structure identical to the transition state of the stepwise case isomerisation (Fig. 1.4). This facts supports the idea that the stepwise pathway of isomerisation of H_2P via the *cis* tautomer is the most appropriate.

The solvent effect on this molecule has been studied by PCM model for simulating toluene and water solvents. However no significant effect was found for either polar or nonpolar solvents, though the first excited triplet state of H_2P seems to be more stabilized with the presence of solvent.

1.2.3 Conclusions

In conclusion, we have studied the ground and triplet state geometry of H_2P by density functional theory calculations. The proton transfer events have been simulated in two different models. The transitions states and intermediate conformations as well as the energy barriers have been calculated. The concerted mechanism for proton shifts was found improbable. The calculations indicate that the isomerisation proceed via *cis* isomer by a stepwise mechanism.

References

1. Hoff, A.J.: Biochim. Biophys. Acta **440**, 765 (1976)
2. Hoff, A.J., Gast, P.: J. Phys. Chem. **83**, 3355 (1979)
3. Hoff, A., Deisenhofer, J.: Phys. Rep. **287**, 1 (1997)
4. Parusel, A.B.J., Grimme, S.: J. Porphyrins Phtalocyanines **5**, 225 (2001)
5. van Dorp, W.G., Soma, M., Kooter, J.A., van der Waals, J.H.: Mol. Phys. **28**, 1551 (1974)
6. van Dorp, W.G., Schoemaker, W.H., Soma, M., van der Waals, J.H.: Mol. Phys. **30**, 1701 (1975)
7. Langhoff, S.R., Davidson, E.R., Gouterman, M., Leenstra, W.R., Kwiram, A.L.: J. Chem. Phys. **62**, 169 (1975)
8. Tsvirko, M.P., Solovjev, K.N., Gradyushko, A.T., Dvornikov, S.S.: Opt. Spectrosc. **38**, 400 (1975)
9. Volker, S., van der Waals, J.H.: Mol. Phys. **32**, 1703 (1976)
10. Gouterman, M.: J. Mol. Spectrosc. **6**, 138 (1961)
11. Langhoff, S.R., Davidson, E.R., Kern, C.W.: J. Chem. Phys. **63**, 4800 (1975)
12. Nakatsuji, H., Hasegawa, J., Hada, M.: J. Chem. Phys. **104**, 2321 (1996)
13. Hasegawa, J., Ozeki, Y., Hada, M., Nakatsuji, H.: J. Phys. Chem. B **102**, 1320 (1998)
14. Serrano-Andres, L., Merchan, M., Rubio, M., Roos, B.O.: Chem. Phys. Lett. **295**, 195 (1998)

15. Sundholm, D.: Chem. Phys. Lett. **302**, 480 (1999)
16. Sundholm, D.: Phys. Chem. Chem. Phys. **2**, 2275 (2000)
17. Sundholm, D.: Chem. Phys. Lett. **317**, 392 (2000)
18. Nguyen, K.A., Day, P.N., Pachter, R.: J. Phys. Chem. A **104**, 4748 (2000)
19. Ohkawa, K., Hada, M., Nakatsuji, H.: J. Porphyrins Phtalocyanines **5**, 256 (2001)
20. McGlynn, S.P., Azumi, T., Kinoshita, M.: Molecular Spectroscopy of the Triplet State. Prentice Hall, Engelwood Cliffs (1969)
21. Tsvirko, M.P., Solovjev, K.N., Gradyushko, A.T., Dvornikov, S.S.: Zh. Prikl. Spektrosk. **20**, 528 (1974)
22. Gouterman, M., Khalil, G.E.: J. Mol. Spectrosc. **53**, 88 (1974)
23. Radziszewski, J.G., Waluk, J., Nepras, M., Michl, J.: J. Phys. Chem. **95**, 1963 (1991)
24. van Dorp, W.G., Schaafsma, T.J., Soma, M., van der Waals, J.H.: Chem. Phys. Lett. **21**, 221 (1973)
25. Clarke, R.H., Connors, R.E., Norris, J.R., Thurnauer, M.C.: J. Am. Chem. Soc. **97**, 7178 (1975)
26. Clarke, R.H., Hobbard, D.R., Leenstra, W.R.: J. Am. Chem. Soc. **101**, 2416 (1979)
27. Clarke, R.H. (ed.): Triplet State ODMR Spectroscopy. Wiley, New York (1982)
28. Connors, R.E., Leenstra, W.R.: In: Clarke, R.H. (ed.) Triplet State ODMR Spectroscopy, p. 258. Wiley, New York (1982)
29. van der Poel, W.A.J., Singel, D.J., Schmidt, J., van der Waals, J.H.: Mol. Phys. **49**, 1017 (1983)
30. Kay, C.W.M., Gromadecki, U., Torring, J.T., Weber, S.: Mol. Phys. **99**, 1413 (2001)
31. Spence, J.D., Lash, T.D.: J. Org. Chem. **65**, 1530 (2000)
32. Tunnell, I., Rinkevicius, Z., Salek, P., Vahtras, O., Minaev, B.F., Helgaker, T., Ågren, H.: J. Chem. Phys. **119**, 11024 (2003)
33. Prabhakar, R., Siegbahn, P., Minaev, B., Ågren, H.: Activation of triplet dioxygen by glucose oxidase: spin-orbit coupling in the superoxide ion. J. Phys. Chem. B **106**, 3742 (2002)
34. Siegbahn, P., Prabhakar, R., Minaev, B.: A theoretical study of the dioxygen activation by glucose oxidase and copper amine oxidase. Biochem. Biophys. Acta **1557**, 205 (2003)
35. Minaev, B.F.: Electronic mechanisms of molecular oxygen bioactivation. Ukrainian J. Biochem. **74**, 11 (2002)
36. Minaev, B.F.: Intensities of spin-forbidden transitions in molecular oxygen and selective heavy atom effects. Int. J. Quantum Chem. **17**, 367 (1980)
37. Minaev, B.F.: About the lifting of spin-forbiddness in reactions of triplet molecular oxygen. Rus. J. Struct. Chem. **23**, 7 (1982)
38. Minaev, B.F.: RIKEN Rev. **44**, 147 (2002)
39. Danovich, D., Shaik, S.: J. Am. Chem. Soc. **119**, 1773 (1997)
40. Yoshizawa, K., Shiota, Y., Yamabe, T.: Chem. Eur. J. **3**, 1160 (1997)
41. Brunold, T.C., Solomon, E.I.: J. Am. Chem. Soc. **121**, 8288 (1999)
42. Brunold, T.C., Solomon, E.I.: J. Am. Chem. Soc. **121**, 8277 (1999)
43. Shaik, S., Filatov, M., Schroder, D., Schwarz, H.: Chem. Eur. J. **4**, 193 (1998)
44. Helgaker, T., Jensen, Aa, H.J., Jørgensen, P., Olsen, J., Ruud, K., Ågren, H., Andersen, T., Bak, K.L., Bakken, V., Christiansen, O., Dahle, P., Dalskov, E.K., Enevoldsen, T., Fernandes, B., Koch, H., Heiberg, H., Hettema, H., Jonsson, D., Kirpekar, S., Kobayashi, R., Mikkelsen, K.V., Norman, P., Packer, M.J., Saue, T., Taylor, P.R., Vahtras, O.: Dalton Release 1.0, An Electronic Structure Program. Manual. University of Oslo, Oslo (1997)
45. Frisch, M.J., Trucks, G.W., Schlegel, H.B., Scuseria, G.E., Robb, M.A., Cheeseman, J.R., Zakrzewski, V.G., Montgomery, J.A., Stratmann, R.E., Jr., Burant, J.C., Dapprich, S., Millam, J.M., Daniels, A.D., Kudin, K.N., Strain, M.C., Farkas, O., Tomasi, J., Barone, V., Cossi, M., Cammi, R., Mennucci, B., Pomelli, C., Adamo, C., Clifford, S., Ochterski, J., Petersson, G.A., Ayala, P.Y., Cui, Q., Morokuma, K., Malick, D.K., Rabuck, A.D., Raghavachari, K., Foresman, J.B., Cioslowski, J., Ortiz, J.V., Baboul, A.G., Stefanov, B.B., Liu, G., Liashenko, A., Piskorz, P., Komaromi, I., Gomperts, R., Martin, R.L., Fox, D.J., Keith, T., Al-Laham, M.A., Peng, C.Y., Nanayakkara, A., Gonzalez, C., Challacombe, M., Gill, P.M.W., Johnson, B., Chen, W., Wong, M.W., Andres, J.L., Gonzalez, C., Head-Gordon, M., Replogle, E.S., Pople, J.A.: Gaussian 98, Revision A.7. Gaussian, Pittsburgh (1998)

46. Woon, D.E., Dunning, T.H.: J. Chem. Phys. **98**, 1358 (1993)
47. Vahtras, O., Minaev, B.F., Loboda, O., Ågren, H., Ruud, K.: Chem. Phys. **279**, 133 (2002)
48. Boorstein, S.A., Gouterman, M.: J. Chem. Phys. **39**, 2443 (1963)
49. Marian, C.M., Wahlgren, U.: Chem. Phys. Lett. **251**, 357 (1996)
50. Barone, V.: In: Chong, D.P. (ed.) Recent Advances in Density Functional Methods (Part 1), p. 287. World Scientific, Singapore (1995)
51. Miertus, S., Tomasi, J.: Chem. Phys. **65**, 239 (1982)
52. Gouterman, M., Adler, A.D., Edwards, L., Dolphin, D.H.: J. Mol. Spectrosc. **38**, 16 (1971)
53. Nakatsuji, H., Takashima, H., Hada, M.: Chem. Phys. Lett. **233**, 95 (1995)
54. Gouterman, M., Yamanashi, B.S., Kwiram, A.L.: J. Chem. Phys. **56**, 4073 (1972)
55. Ågren, H., Vahtras, O., Minaev, B.F.: Adv. Quantum Chem. **27**, 71 (1996)
56. Minaev, B.F., Tunnell, I., Salek, P., Loboda, O., Vahtras, O., Ågren, H.: Mol. Phys. **102**, 1391–1406 (2004)
57. Kay, C.W.M.: J. Am. Chem. Soc. **125**, 13861–13867 (2003)
58. Nakatsuji, H., Hada, M., Tejima, T., Nakajima, T., Sugimoto, M.: Chem. Phys. Lett. **249**, 284 (1996)
59. McConnell, H.M.: J. Chem. Phys. **24**, 460 (1956)
60. Minaev, B.F.: Spin allignement of the triplet state and new methods of the phosphorescence study. Fizika Molecul, Naukova Dumka, Kiev 7, 34 (1979) (in Russian).
61. Ivanov, A.I., Mikhailova, V.A., Feskov, S.V.: A model of spin catalysis in bacterial photosynthetic reaction. Appl. Magn. Reson. **16**, 481 (1999)
62. Minaev, B.F.: Serebrennikov, YuA: Microwave energy transfer between molecular triplet states at the optical spin allignement. Izv. Vyssh. Uchebn. Zaved. Fiz. **7**, 119 (1979)
63. Minaev, B.F., Loboda, O., Vahtras, O., Ruud, K., Ågren, H.: Solvent effects on optically detected magnetic resonance in triplet spin labels. Theor. Chim. Acta **109**, 44 (2003)
64. Lemaistre, J.P., Zewail, A.H.: Chem. Phys. Lett. **68**, 302 (1979)
65. Serebrennikov, Y.A., Minaev, B.F.: Magnetic field effects due to spin-orbit coupling in transient intermediates. Chem. Phys. **114**, 359 (1987)
66. Minaev, B.F.: Spin-orbit interaction in molecules and mechanism of the external magnetic field effect on luminescence. Opt. Spectrosc. (USSR) 44, 148 (1978)
67. Soloviev, K., Zalesskii, I., Kotlo, V., Shkirman, S.: J. Exp. Theor. Phys. Lett. **17**, 332 (1973)
68. Loboda, O., Tunell, I., Minaev, B., Agren, H.: Chem. Phys. **312**, 299 (2005)
69. Becke, A.: J. Chem. Phys. **98**, 5648 (1993)
70. Storm, C., Teklu, Y., Sokoloski, E.: Ann. N.Y. Acad. Sci. 206, 631(1973)
71. Abraham, R., Hawkes, G., Smith, K.: Tetrahedron Lett. **16**, 631 (1973)
72. Schlabach, M., Rumpel, H., Limbach, H.: Angew. Chem. Int. Ed. Engl. **28**, 76 (1989)

Chapter 2
Exohedral Metallofullerenes

2.1 Monoatomic Doping

Fullerene-based materials have attracted considerable interest since the discovery of C_{60}. A promising area of research concerns metal–fullerene interactions and their application to advanced nano materials, with potential use in optical and switching devices, as photoconductors, and for hydrogen storage. Moreover, transition metal complexes of fullerenes show catalytic activity in homogeneous hydrogenation of acetylenic alcohols [1] and hydroformylation of alkenes [2]. In heterogeneous catalysis, exohedral metallofullerenes are found to promote hydrogenation of olefins and acetylenes [3, 4] as well as reduction of carbon monoxide to methane [5, 6].

The first exohedral metallofullerenes to be synthesized and isolated were η^2 complexes of $C_{60}Pt(PPh_3)_2$ and $C_{60}Pd(PPh_3)_2$ [7–9]. This work was later extended to monoatomic coordination by a whole range of metals including Ti, Nb, Re, Fe, Ru and Ni, and also multiple metal doping [4, 10–13].

Although geometry optimisation and calculation of electronic properties of metallofullerenes is time consuming because of the size of the system, the development of density functional theory (DFT) methods has made it possible to obtain accurate theoretical results for these systems. There are a limited number of theoretical studies published on exohedral transition metal complexes of fullerenes [14–16]. Most of the attention has been on the interaction of fullerenes with $M(PH_3)_2$ where M is a transition metal atom. The choice of metals was restricted to a single group in the periodical table, and to our knowledge there are no systematic studies of geometry and bonding in metal–fullerene complexes reported so far.

Except for one early report [17] on the bond dissociation energies for the naked nickel–fullerene complex, there is little information available about relative stabilities and dissociation energies of bare metal–fullerene complexes. Why does some metals form strong bonds to fullerenes, while the exohedral metallofullerene CoC_{60} has never been observed? The aim of our study is to shed light on this and related issues, by exploring which factors play important roles in metal–fullerene interactions and hence determine the bond strengths. To this end we study the interaction between

O. Loboda, *Quantum-Chemical Studies on Porphyrins, Fullerenes and Carbon Nanostructures*, Carbon Nanostructures, DOI: 10.1007/978-3-642-31845-0_2,

C_{60} and naked transition metal atoms of group 9 and 10, by means of DFT. Various possible bonding modes of C_{60} to metal atom are explored, including coordination with different hapticity. The M–C_{60} bonds are analyzed in terms of electron donation and back donation and repulsive polarization.

2.1.1 Method of Calculations

The geometries of metallofullerenes were optimized using density functional theory in terms of the OLYP functional [18] in conjugation with analytic gradient techniques.

The carbon, hydrogen and phosphor elements were described by Dunning-Huzinaga double-ζ (DZ) basis set [19, 20]. The corresponding contraction for C and H is (9s, 5p)/[4s, 2p], (4s)/[2s], while the phosphorus (11s, 7p)/[6s, 4p] basis set was augmented by polarization (*d*) function [21]. For the metal atoms the Stuttgart relativistic, small-core ECP [22, 23] basis set was used with cores of 10,28 and $60e^-$ for the first, second and third row respectively, whereas the valence electrons were described by (8s, 7p, 6d)/[6s, 5p, 3d]-contracted basis set. The energy as well as other properties were reevaluated in subsequent single-point calculations using the B3LYP functional [24] and somewhat extended basis sets. The carbon basis was extended with single sets of diffuse *sp* functions and a set of polarization (*d*) functions [19, 20]. A set of diffuse *p* functions was added to phosphorus, with the orbital exponent chosen to form a geometric series with the two most diffuse functions already present in the primitive basis set. Bond dissociation energies (BDE) for M–C_{60} were calculated from the energies of equilibrium structures of the complexes and their constituent fragments according to: $\mathrm{BDE(M-C_{60})} = \mathrm{E(C_{60})} + \mathrm{E(M)} - \mathrm{E(MC_{60})}$. Basis set superposition errors (BSSE) were estimated by the counterpoise correction method [25].

2.1.1.1 Atomic States

While calculation of atomic energies of closed-shell atoms is usually unproblematic, treatment of open-shell atoms with high multiplicity is often complicated when DFT is invoked. Contemporary DFT implementations rely on a single-determinant description of the non-interacting reference system, which implies that one can not apply linear combinations of determinants in order to describe a multiplet system correctly [26]. The course to take is to select a single-determinant non-interacting Kohn-Sham reference system that defines the proper values of the conserved quantum numbers. For the present metal system, the problem reduces to selecting the occupation of real, cartesian *nd*-orbitals that corresponds to a single determinant representation of the desired atomic term with correct angular momentum and spin symmetry. This has been addressed in Refs. [26, 27], and the guideline is to accept only solutions with the correct configuration within integer *d*-orbital occupations. To fulfill this requirement one frequently needs to make use of symmetry restrictions given by non-Abelian point groups. For instance, octahedral symmetry with inversion (O_h) prevents *s*/*d* mixing since these orbitals belong to the different irreducible representations. In most difficult cases it is helpful to use of atomic orbitals obtained from restricted open-shell procedures as initial guesses for unrestricted calculation. Finally one should carefully check that the initially defined orbital assignment of electrons persists until convergence of the self-consistent field. Hay has made a convenient summary of proper occupations for some important *d*-states [27]. Occupations for the d^7, d^8 and d^9 states that are relevant to our systems are the following:

$(d^7s^2)^4F\ (d_{z^2})^2(d_{x^2-y^2})^2(d_{xy})^1(d_{xz})^1(d_{yz})^1$–Co, Ir
$(d^8s^1)^4F\ (d_{z^2})^1(d_{x^2-y^2})^1(d_{xy})^2(d_{xz})^2(d_{yz})^2$–Rh
$(d^9s^1)^3D\ (d_{z^2})^1(d_{x^2-y^2})^2(d_{xy})^2(d_{xz})^2(d_{yz})^2$–Ni,Pt.

It should be noted that for the nickel atom the ground state is often stated to be the ^3F term. Experimentally, the splitting between the ^3F and ^3D states is only 0.03 eV, with $(d^9s^1)^3D$ being the lower one [28].

2.1.1.2 Bonding Analysis

Atomic charges and orbital populations has been computed by means of Natural Bond Orbital analysis [29]. In order to estimate quantitatively the importance of ligand $\longrightarrow$ metal σ-donation and metal $\longrightarrow$ ligand π-back-donation to interaction energies, we used the Charge Decomposition Analysis (CDA) [30] method of Frenking and coworkers [30] as implemented in the AOMIX program [31, 32].

In this method the wave function of any given AB complex is expressed as a linear combination of the canonical molecular orbitals (Φ_μ) of the fragments A and B. The orbital contribution of the fragments to the wave functions of the complex accordingly to CDA method consists of the following terms

- charge donation σ from fragment A to fragment B (mixing of the occupied MOs of A fragment and the vacant MOs of fragment B):

$$\sigma_i = \sum_{k}^{occ,A} \sum_{n}^{vac,B} m_i c_{ki} c_{ni} \langle \Phi_k | \Phi_n \rangle \tag{2.1}$$

  where m_i is the occupation number of ith molecular orbital of AB complex, the summation of σ_i overall molecular orbitals gives the total charge donation (σ);
- back-donation π (mixing of the occupied MOs of B fragment and the vacant MOs of fragment A):

$$\pi_i = \sum_{l}^{occ,B} \sum_{m}^{vac,A} m_i c_{li} c_{mi} \langle \Phi_l | \Phi_m \rangle \tag{2.2}$$

- repulsive polarization r (closed-shell interaction of the occupied MOs of fragment A and B):

$$r_i = \sum_{k}^{occ,A} \sum_{m}^{occ,B} m_i c_{ki} c_{mi} \langle \Phi_k | \Phi_m \rangle \tag{2.3}$$

and the rest term Δ (mixing of vacant MOs) which can be used as indicator of bond covalency rather than a donor-acceptor nature of bond if $\Delta \gg 0$. The former term, repulsion has the negative sign and describes the amount of electronic charge removed from overlap of the occupied orbitals of the fragments into the nonoverlapping regions [30].

2.1.2 Results

2.1.2.1 Geometrical Parameters

The C_{60} Buckminsterfullerene has only two symmetry-unique bonds, corresponding to the junction of two six-membered rings (6–6′) and a six- and a five-membered ring (6–5). We have optimized the geometry of C_{60} under constraints of I_h symmetry, to obtain carbon–carbon bond lengths of 1.411 and 1.465 Å for the two unique distances. These values are in good agreement with data from electron diffraction studies [33], which gave values of 1.401 and 1.458 Å for these bond lengths. The curvature of this fullerene molecule may be characterized by the pyramidalization angle (δ), i.e. the angle between the projection of a specific C–C bond and a plane containing one of the designated carbons and its two adjacent atoms, cf Fig. 2.1. In our optimized C_{60} structure, the pyramidalization angle is in full agreement with the experimental value of 31°.

In order to validate the OLYP functional for the present use, we optimized C_{60}, as well as $Pd(PH_3)_2$- and $Pt(PH_3)_2$-C_{60} using first OLYP and then the B3LYP energy functional. Switching from OLYP to B3LYP induces only very minor changes in the bond distances of C_{60} (1.404 and 1.464 Å). For the metallofullerenes, optimized geometrical parameters are listed together with relevant experimental data in the Table 2.1. The two DFT methods leads to very similar results for the metal–carbon and carbon–carbon distances, whereas the metal–phosphorus distance systematically comes out longer at the B3LYP level of theory, by some 0.03 Å. From Table 2.1, there are significant differences between our geometry parameters and those obtained using the BP86 functional in Ref. [34]. Moreover, similar discrepancies are found between results obtained with OLYP and BP86 functionals for the corresponding ethylene complexes, cf the lower half of Table 2.1. This suggests that the difference is due to the choice of energy functional, since the two sets of BP86 results were obtained using different basis sets. In order to decide which is the better functional to use for this kind of study, it is desirable to make comparison to experimental structures. However, experimentally the triphenylphosphine ligand is used rather than phosphine itself, which complicates a direct comparison between theory and experiment. A comparison is further complicated by packing effects on the experimental X-ray

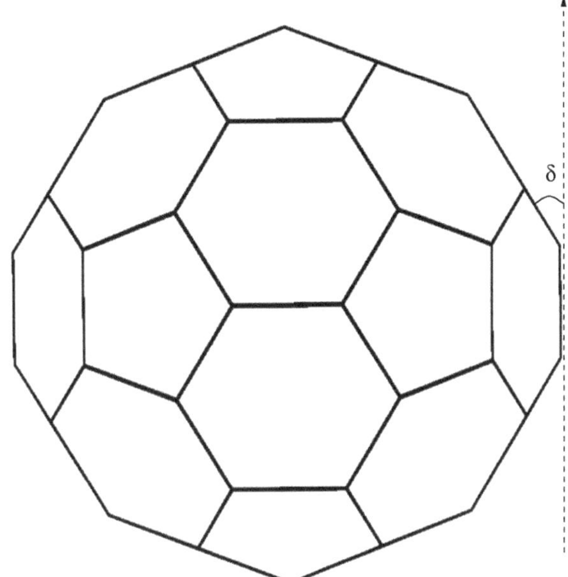

Fig. 2.1 Schematic representation of C_{60} and pyramidalization angle δ

structures, indicated by presence of inequivalent complexes in the unit cell, as well as distortion due to inclusion of solvent molecules in the crystal. The data included in Table 2.1 does not give a decisive edge to any of the functionals BP86 and OLYP, but rather serves to illustrate the level of uncertainty attached to both the experimental and theoretical structures.

Upon η^2-coordination of $M(PH_3)_2$ to a (6–6′) junction at C_{60}, interacting C–C bond extends to 1.489 Å in $Pd(PH_3)_2\eta^2$-C_{60} and to 1.516 Å in $Pt(PH_3)_2\eta^2$-C_{60}, quoting numbers obtained from OLYP. In the calculations, platinum approaches closer to the fullerene than does palladium, cf M–C distances of 2.14 and 2.16 Å, although the opposite is reported for the X-ray structure of the analogous triphenylphosphine complexes. The error bars are quite large, however, and prevents from drawing firm conclusions on this aspect.

2.1.2.2 Accuracy of OLYP

Let us consider accuracy of OLYP functional on the example of $Pt(PH_3)_2\eta^2$-C_{60}. This complex has been studied by different functionals such as B3LYP [35], BP86 [34, 36]. OLYP functional gives the best value for P–Pt–P angle (103.1°) versus BP86 [34] (107.4°) which is in a good agreement with the experiment [7] (102.4°). B3LYP value for P–Pt–P angle is in intermediate position (104.1°) between OLYP and BP86. The metal–phosphorus bond length has the same accuracy (2.297 Å) in OLYP and BP86 [36] calculations and quite close to the averaged experimental [7]

Table 2.1 Geometrical parameters as optimized for $M(PH_3)_2(\eta^2\text{-}C_{60})$ and $M(PH_3)_2(\eta^2\text{-}C_2H_4)$, M = Pd and Pt, and compared experimental data for the corresponding triphenylphosphine complexes

	rM–C	rC–C	rM–P	∠CMC	∠PMP	δ^a
$Pd(PH_3)_2(\eta^2\text{-}C_{60})$						
OLYP	2.163	1.489	2.336	40.3	105.3	39.5
B3LYP	2.162	1.482	2.364	40.1	106.9	39.6
BP86[b]	2.180	1.464	2.378	39.2	111.0	38.8
$Pd(PPh_3)_2(\eta^2\text{-}C_{60})$ Exp[c]	2.123 [14]	1.447 [25]	2.330 [6]	40.2	109.7	40 [2]
	2.086 [16]		2.315 [5]			38 [2]
$Pt(PH_3)_2(\eta^2\text{-}C_{60})$						
OLYP	2.143	1.516	2.297	41.4	103.1	41.9
B3LYP	2.147	1.511	2.324	41.2	104.1	41.5
BP86[a]	2.103	1.505	2.289	41.9	107.4	41.8
$Pt(PPh_3)_2(\eta^2\text{-}C_{60})$Exp[d]	2.145 [24]	1.502 [30]	2.253 [7]	41.3 [8]	102.4 [2]	38 [2]
	2.115 [23]		2.303 [7]			44 [2]
$Pd(PH_3)_2(\eta^2\text{-}C_2H_4)$ OLYP	2.182	1.424	2.319	38.1	107.2	(19.3)
$Pd(PH_3)_2(\eta^2\text{-}C_2H_4)$ BP86[e]	2.195	1.401	2.312	37.2	112.5	–
$Pt(PH_3)_2(\eta^2\text{-}C_2H_4)$ OLYP	2.147	1.453	2.291	39.6	104.4	(24.5)
$Pt(PH_3)_2(\eta^2\text{-}C_2H_4)$ BP86[e]	2.152	1.431	2.286	39.0	108.7	–
$Pt(PPh_3)_2(\eta^2\text{-}C_2H_4)$ Exp[f]	2.106 [4]	1.434 [2]	2.270 [4]	39.7 [4]	111.60	
	2.116 [9]		2.265 [4]			

Bond lengths in Å, angles in degrees

[a]Pyramidalization angle as defined in the text. The definition is modified as angle between C–C projection and CH_2 plane when applied to the ethylene complexes

[b]Ref. [34]

[c]Ref. [9]

[d]Ref. [7]

[e]Ref. [52]

[f]Ref. [53]

value 2.278 Å. The calculated B3LYP value (2.324 Å) for Pt–P bond stands up far from the OLYP and BP86 results. The B3LYP Pt–P bond length (2.418 Å) reported in Ref. [35] overestimates the metal–phosphorus bond length. The fact is the Pt–P bond is very sensitive to the chosen basis set on phosphorus. We found out that inclusion of polarization function on phosphorus improves the value for Pt–P bond length significantly. The calculated BP86 [34] value 1.505 Å for interacting C–C distance agrees well with the experiment [7] 1.502 Å. The other BP86 [36] value 1.495 Å is also close to the experiment data. The corresponding bond distance obtained by OLYP functional 1.516 Å is somewhat longer. As one can see there is no unambiguous solution in the choice of functional. Each of the discussed functionals has the tendency to overestimate one parameter and underestimate the other. The results from OLYP geometry optimisation display good agreement with the experiment and comparable to accuracy of BP86 and B3LYP functional calculations.

2.1.2.3 Pd versus Pt in the Phosphine Complexes

It is noteworthy that Pd–C bond length in ethylene and fullerene complexes is longer than the Pt–C distance. This trend has been also observed at BP86 [34] level of theory. Kameno and coworkers [35] provide explanation to this trend which is based on the fact that the binding energy of $Pd(PH_3)_2–C_{60}$ is smaller than that of Pt analogue. Indeed, we found similar consistency between the bond dissociation energy and metal–carbon distance in Pt and Pd compounds. However this can not be applied for Ni and Co analogues where the metal–carbon bond is shortest while the binding energy is lowest, but this will be the subject of our discussion below.

Also it is known that the inclusion of relativistic effects on heavy atoms shortens the metal–ligand bond distance in π-complexes significantly [37]. This being the case, the relativistic contraction equalizes Pt–C and Pd–C bond distances in $M(PH_3)_2C_2H_4$ complexes whereas nonrelativistic order gives Pd–C > Pt–C.

2.1.2.4 Ethylene versus C_{60}

The coordination of metal–phosphine ligand to ethylene results in deviation from planar structure. Calculated angular deformation between the C–C bond and CH_2 plane is 19.3° for $Pd(PH_3)_2C_2H_4$ and 24.5° for Pt analogue. Likewise the metal–ethylene bonding, in the case of fullerene the pyramidalization angle is increased upon coordination to metal atom from 31.7° in free C_{60} to 39.5° and 41.9° in Pd, Pt-fullerene derivatives (look Table 2.1). This effect is remarkable on the carbons at the reaction site. However at non-interacting region of C_{60} framework no significant perturbations in geometry have been found. Therefore exohedral interaction of metal atom with fullerene molecule is local phenomena.

In general, the geometry of $M(PH_3)_2(\eta^2C_{60})$ resembles precursor ethylene parent adduct $M(PH_3)_2(\eta^2\text{-}C_2H_4)$ except the fact that in fullerene complexes M–P and C–C bonds as a rule are slightly longer than in ethylene compounds (Table 2.1). Moreover the phosphorus in phosphine complexes prefer to stay in the same plane with the metal atom and interacting carbons of ethylene or fullerene molecule.

2.1.2.5 Binding Sites

We investigated the relative stability of fullerene complexes with different locations of palladium atom. For this purpose we optimised PdC_{60} structures with different hapticity, where metal atom is bound over pentagonal (η^5), hexagonal(η^6) rings and above two carbon atoms(η^2) of (6–6′) and (6–5) ring junction look Fig. 2.2. The cohesive energies corresponding to formation of palladium–fullerene bond are tabulated in Table 2.2. The most stable structure was found η^2 complex of metallofullerene in which palladium binds to carbons at the fusion of two six-membered rings (6–6′). This is in accord with the experiment since it is well-established that (6–6′) bonds of C_{60} is shorter than the (6–5) bonds and have the most double-bond character [38–40].

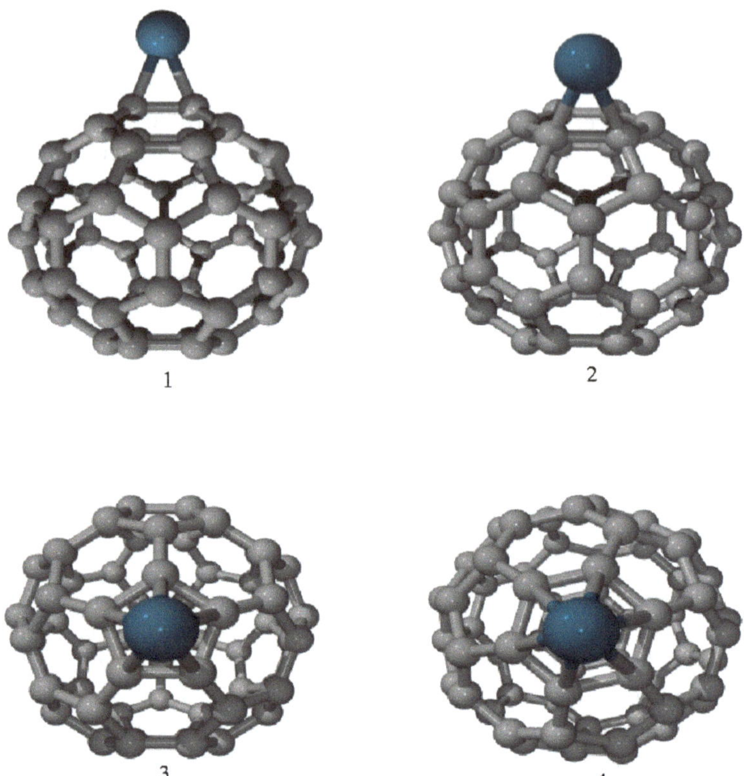

Fig. 2.2 Optimised structures of PdC_{60}: *1* η^2(6–6), *2* η^2(5–6), *3* η^5, *4* η^6

Table 2.2 Palladium bond dissociation energies (BDE) at different sites of C_{60} compound (kcal/mol), look Fig. 2.2

	η^2(6–6)	η^2(5–6)	η^5	η^6
BDE[a]	33.9	27.3	15.4	10.2
BSSE	1.9	1.9	2.8	2.5

[a]BSSE is not included

After we determined the most stable structure for PdC_{60} complex we expanded our investigations on stability of metallofullerenes for transition metals of group 9 and 10. Metal–fullrene bond dissociation energies, metal–carbon distances and pyramidalized angle are presented in Table 2.3. The coordination energies of naked metal atoms to C_{60} is found to increase down the groups, with the highest bond dissociation energy calculated for Pt–η^2C_{60}. Complexation of metal atom to C_{60} bond results in elongated C–C bonds at the site of metal coordination compared to those in C_{60} (Tables 2.1, 2.3). The electron-rich metal substance lowers the electron affinity of C_{60} by increasing electron population. Thus the resulting negative charge

Table 2.3 The bond dissociation energies ($M–C_{60}$) (kcal/mol), metal–carbon, carbon-carbon distances(Å) and pyramidalization angle of MC_{60}

M	Molecular state	BDE[a]	BSSE	R^{M-C}	$R^{C=C}$	δ
Co	$^2A_1$	7.7 (5.2)	2.6	1.951	1.514	40.4
Rh	$^2A_1$	37.9 (33.3)	1.9	2.073	1.510	40.1
Ir	$^2A_1$	41.7 (44.8)	2.1	2.054	1.531	41.3
Ni	$^1A_1$	23.3 (18.6)	2.5	1.912	1.508	39.8
Pd	$^1A_1$	33.9 (32.3)	1.9	2.114	1.485	38.2
Pt	$^1A_1$	49.7 (52.3)	2.0	2.060	1.516	40.0

[a]BSSE is not included; the bond dissociation energy of $M–C_2H_4$ is in parenthesis

Table 2.4 NBO charges of metal and interacting carbons, dipole moment (Debye) and HOMO-LUMO(ΔE) energy gap (eV) in the $M–C_{60}$ complexes

Charge, e	Ni	Pd	Pt	Co	Rh	Ir
Metal	0.46	0.30	0.20	0.62	0.37	0.26
Carbon	−0.10	−0.050	−0.01	−0.15	−0.06	−0.02
μ	4.46	3.14	2.01	6.10	4.39	3.04
ΔE	2.34	2.48	2.49	1.96	2.43	2.44

Table 2.5 Bond dissociation energies BDE (kcal/mol) and dipole moment μ (Debye)

	BDE[a]	BSSE	μ		BDE[a]	BSSE	μ
Pt–C2H4	52.25	0.46	0.09	Pd–C2H4	32.27	0.40	0.60
Pt(PH3)2–C2H4	13.95	1.37	2.33	Pd(PH3)2–C2H4	11.55	1.16	1.87
$Pt(PH_3)_2–C_{60}$	17.83	4.81	9.17	$Pd(PH_3)_2–C_{60}$	16.55	4.32	8.05

[a]BSSE is not included

is mainly confined to the carbon atoms directly engaged in the interaction with the metal. In line with this, the dipole moments are moderate in magnitude despite the large size of these complexes, decreasing in groups from 4.5 D in NiC_{60} to 2.0 D in PtC_{60} and from 6.1 D in CoC_{60} to 3.0 D in IrC_{60} (Table 2.4). Addition of phosphine ligands to the metal atom increases the dipole moment significantly, but decreases metal–fullerene BDE look Table 2.5.

2.1.3 Discussion

It has been shown [41, 42] that partial donation of the π-electrons from alkene to an empty σ-orbital of the metal weakens the π-bond of the unsaturated hydrocarbon and therefore lowers π^* energy, which makes electrons easier accepted from a back-donating d-orbital of the metal atom. Transfer of electrons from the metal d-orbital to the antibonding π^* orbital increases the energy of the latter. As the consequence the degree of forming of metal–alkene bond should be reflected in HOMO-LUMO energy gap of the complex. According to this mechanism the most strong M–C_{60} bond should have the metallofullerene complex with the largest HOMO-LUMO gap. In the complexes with various metal atoms the amount of charge transfer from fullerene to the metal as well as back-donation from metal to fullerene will be depend on the metal's ns and nd orbital energy. The lower ns–nd energy gap the better sd hybridisation and the stronger metal–fullerene bond is formed. The similar radial distribution of s and d orbitals facilitates sd hybridisation and therefore favors the interaction with the corresponding carbon orbitals.

Indeed inspection of the HOMO-LUMO(ΔE) gap in M–C_{60} complexes revealed correlation with the bond dissociation energy (BDE). The smallest energy gaps 1.96 and 2.34 eV were found for the top of group 9 and 10, namely Co,Ni, which are the most weakly bound to the fullerene (Table 2.4). Meanwhile the most stable compounds are Ir and Pt complexes. They have the largest HOMO-LUMO gap and the largest BDE in the groups. In fact the second- and third-row metals usually have stronger metal–carbon bonds than the first-row metals [43, 44]. The 5s and 4d orbitals of the second row metals contract much better than the 4s and 3d orbitals of the first-row metals.

Likewise, energy gap between s and d orbitals for second and third-row metal atoms is effectively smaller than for the first-row metal atoms [45]. The strength of the metal–alkene bond depends on the degree of spatial overlap of atomic orbitals which occur during approaching of metal atom to the alkene. In order to achieve the proper overlap the sd/sp hybridisation should occur in metal atom. The hybridisation relieves the repulsion between the ns orbital of metal atom and π orbital of alkene due to the fact that one of the hybrid orbital accumulates the electronic charge in nonbonding region [46]. At the same time the other hybrid orbital becomes the acceptor orbital, which is oriented along the coordination plane.

We used radius and energy values of s, d orbitals as variables for prediction of BDE (look Fig. 2.3). The correlation coefficient (r) is equal to 0.91 and defined as:

$$r = \frac{C_{ij}}{\sqrt{C_{ii}C_{jj}}} \tag{2.4}$$

where C_{ij} is covariance between calculated and predicted data sets and C_{ii}, C_{jj} are standard deviations.

The BDEs based on this model is in qualitative agreement with calculated BDEs however it overestimates energy values for Co and underestimates for Ni i.e. it deviates in the cases where promotion energy should be taken into account.

From the Table 2.4 one can notice that Co has the lowest BDE. This is also supported by experiment since the neutral Co atom show no reaction neither with ethene [47] nor with fullerene [13].

The low bond dissociation energy of cobalt–fullerene(ethene) compounds can be explained in terms of s–d hybridisation mentioned above, loss of exchange and promotion energies. For the beginning it should be noted that atomic ground state of neutral cobalt is $3d^7 4s^2$. According to ab initio calculations [48, 49] $M(d^{n-2}s^2)$ +

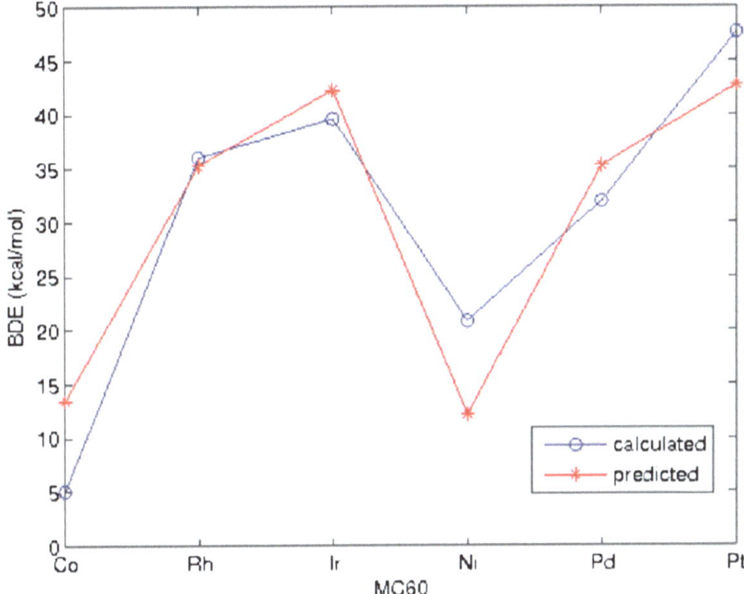

Fig. 2.3 Linear regression model for BDE of MC_{60}. *Blue line* is calculated BDEs; *red line* is predicted BDE based on radii extension and energy gap of s, d orbitals

C_2H_4 asymptotes exhibit only repulsive potentials. Essential metal–alkene binding arises only from attractive $d^{n-1}s$ asymptotes. In order to reach excited $d^{n-1}s$ state cobalt atom should obtain promotion energy at least 10 kcal/mol [28]. Our own calculations determine 13 kcal/mol promotion energy to $d^{n-1}s$ state of low-spin for the cobalt and only 3 kcal/mol for the nickel atom. Only in such bond prepared state the metal atom can hybridise and form the bond with the alkene. Contribution to low BDE of Co atom gives also large s/d energy gap and large ratio between radii extension of these orbitals. For the other complexes with the strong metal–fullerene bond these parameters have significantly smaller values.

Considering M–C_{60} interaction in terms of Dewar–Chatt–Duncanson model [50, 51] one should study in a couples σ-donation of π-electrons to the metal and a π-bond back-donation from metal to the lowest vacant orbitals of interacting carbon (6–6′) bond.

Table 2.6 summarizes total donation, back-donation and repulsion terms for metallofullerenes and for the M–C_2H_4 parent adduct. Both ethylene and fullerene complexes have common peculiarities: σ donation decreases downwards the group, while π back-donation has the opposite trend and increases in the group with the increasing of nuclear charge of the metal. For the late transition metals studied here, except of the first-row transition metals (Co, Ni), the metal–C_{60}/C_2H_4 bond is dominated by back-donation. We found that the M–C_{60}/C_2H_4 bond strength correlates with the donation and back-donation (look Tables 2.3, 2.6). BDE increases with the

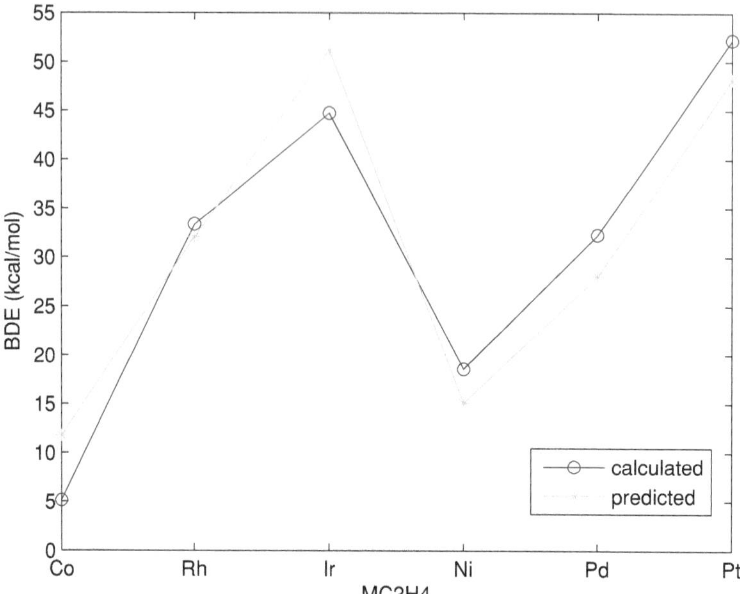

Fig. 2.4 Linear regression model for BDE of MC_2H_4. *Blue line* is calculated BDEs; *red line* is predicted BDE based on π, σ variables

increasing of π back-donation and decreasing of σ donation. For example the largest values for back-donation 0.42 e exhibits the Ir and Pt-fullerene complexes, which have also the largest BDE 42 and 50 kcal/mol respectively. We used multivariate method of analysis for processing of BDE, σ and π data. The correlation diagrams are presented in Fig. 2.4 for MC_2H_4 and in Fig. 2.5 for MC_{60}. Obtained correlation coefficient of regression model based on two variables, σ and π is 0.99 for PdC_{60}. The corresponding correlation coefficient for PdC_2H_4 is 0.95.

This type of behavior in BDE and charge transfer between fullerene and metal atoms can be understood on the ground of the s and d orbital energies of the metal atom. As we mentioned above the size and energy of s, d orbitals play an important role in the metal–fullerene bonding. It is clear that if the energy of d orbital rises then the back donation will increase too, and with the lowering of s orbital energy one can expect decreasing of σ donation. Since down the group the d orbital energy increases the same trends one can observe in back-donation process, which leads to the corresponding trend in BDE. And the results from the Tables 2.3, 2.6 supports this conclusion. Inspection of the repulsion term shows significant closed-shell repulsion for the Rh and Pd species. The reduced repulsive polarization in metallofullerene complexes in comparison to the metal–ethylene compounds is assumed due to the reduction of electron occupancy on interacting carbon (6–6′) bond of fullerene molecule. The NBO population of this bond in pure fullerene is 1.64 whereas in ethylene

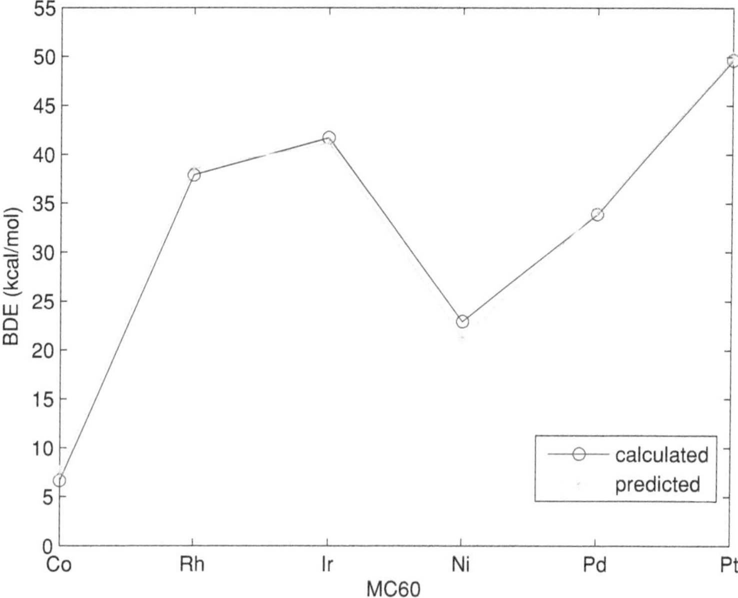

Fig. 2.5 Linear regression model for BDE of MC_{60}. *Blue line* is calculated BDEs; *red line* is predicted BDE based on π, σ variables

Table 2.6 Charge decomposition analysis of MC_{60}[a] and MC_2H_4: donation (σ), back-donation (π) and repulsion (r)

MC_{60}	σ	π	r	MC_2H_4	σ	π	r
CoC_{60}	0.33	0.20	−0.11	CoC_2H_4	0.42	0.22	−0.21
RhC_{60}	0.22	0.37	−40.20	RhC_2H_4	0.20	0.32	−0.34
IrC_{60}	0.12	0.42	−0.17	IrC_2H_4	0.08	0.41	−0.14
NiC_{60}	0.32	0.26	−0.10	NiC_2H_4	0.31	0.24	−0.02
PdC_{60}	0.25	0.34	−0.18	PdC_2H_4	0.25	0.30	−0.35
PtC_{60}	0.21	0.42	−0.15	PtC_2H_4	0.22	0.39	−0.24

[a]CDA based on DZP basis set calculations (In order to fulfil the requirement of AOMIX program 'Number of canonical orbitals = number of basis functions' and to avoid linear dependency problem we had to run B3LYP single point calculations without diffuse basis function on carbon. Obtained BDE are very much similar to the previous and does not differ from them by more than 0.3 kcal/mol)

molecule the population of C–C bond is close to the double occupancy. Thus the repulsive polarization in MC_{60} is less than metal–ethylene four electron repulsion.

It should be noted that the difference between the amount of donation and back-donation is not equal to the charge transfer between metal and fullerene. This is because the terms donation and back-donation according to CDA method [30] do not include only charge transfer interactions but rather describe an overall reorganization of electronic density. The stronger electronic polarization the greater deviation between the difference of donation and back-donation and the net charge transfer is expected [31, 32].

2.1.4 Conclusions

The different types of structures on the ground state of selected MC_{60}, have in the present work been discussed together with the interpretation of bond dissociation energies. In general the site above a 6–6′ ring junction leads to an attractive interaction.

Two DFT methods, namely OLYP and B3LYP, have been compared via geometrical parameters and stability of various exohedral metallofullerene structures.

The M–C_{60} bonds are analyzed in terms of Charge Decomposition Analysis giving quantitative estimates of electron donation, back donation and repulsive polarization. The π back-donation has the dominant contribution to the metal–fullerene bond for the second and third row metal–fullerene bond for the second and third row metal MC_{60}/C_2H_4 complexes. We found and interpreted the correlation between BDE and amount of π back-donation and σ donation.

The weak Co–C_{60} bond is due to the large promotion energy of neutral cobalt atom to the excited bond-prepared state.

The charge decomposition analysis based on linear combination of fragment molecular orbitals is proved to be very useful tool for studying interaction between molecular fragments in terms of donation, back-donation and polarization.

2.2 Multiple Doping

2.2.1 Overview

A number of compounds in which fullerene cages are exohedrally and multiply doped with bare alkali [54], alkaline earth [54] or transition metal atoms [54–59] have been synthesized and isolated. Examples of the latter include M_nC_{60}, M = Ti, Nb, Re, Fe, Ru, Ni, Pd and Pt. For M_nC_{60} n > 1 only Ni_nC_{60} has been investigated theoretically [60, 61]. Most of the experimental attention has been directed toward palladium-and platinum-doped C_{60} [54, 56, 59, 62], and by varying the reaction conditions, Pd_nC_{60} with $n = 1 - 7$ may be synthesized [63]. Difficulties in obtaining single crystals of the required size and quality [54] have meant that only limited structural information is available for these compounds. The monosubstituted fullerene PdC_{60} is believed to exist as a linear polymer, with palladium atoms alternating with C_{60} units [55]. Structures with higher degree of doping, i.e. $n > 1$ are assumed to have crosslinks between monomer chains [55].

Palladium–fullerene complexes exhibit catalytic activity with respect to hydrogenation of unsaturated hydrocarbons [55, 63]. It has been suggested that there are two types of metal atoms in Pd_nC_{60}, and that palladium atoms bridging neighboring fullerene units are catalytically inactive [55, 63]. The palladium particles responsible for the catalytic activity are believed to be "excessive" metal atoms adsorbed on the surface of C_{60} [55, 63]. A clear picture of how multiple metal atoms arrange on the fullerene surface is, however, not available, and it is not known whether the "excessive" metal atoms adsorb individually or form metal clusters on the fullerene surface.

The aim of the present contribution is to identify the preferred structures of C_{60} with two or more bare palladium atoms adsorbed, as well as to assess the associated Pd-fullerene bond energies. To this end we use density functional theory to optimize the geometry of the molecules and to compute the binding energy of palladium atoms at different sites.

Formation of equivalent bonds in metal–ligand bonding is referred to as hapticity (η), and in the case of exohedral metallofullerenes, single atoms may in principle explore the range from one to six for η, depending on the site of coordination to the fullerene. To our knowledge, only η^2-substituted exohedral metallofullerenes have been observed so far and in the present work we focus on this coordination mode.

The study disregards interactions between fullerene cages, and as such pertains to a gas or dilute solution of these compounds, and to the solid phase only to the extent that metal atoms not taking part in fullerene bridging are taken into consideration.

2.2.2 *Computational Details*

Metallofullerene geometries were optimized using analytic gradient techniques and the OLYP density functional. For Pd, the Stuttgart relativistic $28e^-$ small core ECP [64] was used in conjunction with a (8s, 7p, 6d)/[6s, 5p, 3d]-contracted valence basis set. Properties were obtained in subsequent single-point energy evaluations using the B3LYP functional with a carbon atom basis set that was extended with diffuse and polarization functions compared to that applied in the geometry optimizations. Basis set superposition errors (BSSE) were estimated by the counterpoise correction method. Further details and references can be found in Ref. [65].

2.2.3 *Results and Discussion*

A single palladium atom binds preferentially in η^2-mode to the junction between two hexagons (6–6′)onC_{60}. In Fig. 2.7 we systematically add palladium atoms to C_{60} at sites that allow for this preferred binding mode while avoiding metal–metal interactions as far as possible. Complexation of a Pd atom induces elongation of the interacting carbon–carbon bond from 1.411 Å in pure C_{60} (the corresponding bond

Table 2.7 Bond dissociation energies (BSSE corrected) of $Pd_{n-1}(\eta^2\text{-}C_{60})\text{-}Pd$, n = 1–6

n	1	2	3	4	5	6
BDE (kcal/mol)	32.0	31.9	31.8	31.6	31.6	31.7

distance from electron diffraction is 1.401 Å [66]) to 1.482–1.485 Å in the metallofullerenes. Moreover, the atomic charge and orbital occupancy of the palladium atom first added is only weakly affected by the additional metal atoms. For instance, its NBO charge drops from 0.30e in the monosubstituted fullerene to 0.27e in Pd_6C_{60}.

The interaction between transition metal atoms (M) and fullerenes is often characterized in terms of the *Dewar–Chatt–Duncanson* model [67, 68], which implies electron donation from bonding orbitals at the organic ligand into unoccupied d orbitals of σ symmetry at the metal as well as back donation from metal d-orbitals of π symmetry into vacant π*. Molecular orbitals at the interacting carbon–carbon double bond. We find that this picture describes the η^2-coordinated Pd atom well at a (6–6′) junction. The occupancy of the donating Pd d_π-orbital remains essentially constant at 1.64 e as additional metal atoms are adsorbed.

In line with the results from the population analysis, the bond dissociation energy (BDE) of each Pd atom remains almost constant, at 32 kcal/mol, for up to 6 added metal atoms (Table 2.7). Evidently, the binding of single metal atoms is a local phenomenon, and only small perturbations propagate through the delocalized π system to neighboring sites. Moreover, the fullerene appears to be a very soft ligand, meaning that its electron-accepting capacity is very slowly saturated, as long as the adsorbed metal atoms are evenly distributed. At this point it is interesting to explore whether the adsorbed Pd atoms have a tendency to cluster together at the surface, and, moreover, whether such a metal cluster would desorb from the fullerene once it is formed. We investigated the relative stability of different disubstituted exohedral metallofullerenes by optimizing structures (**1–5**, Fig. 2.7) with different locations of the second Pd atom relative to the palladium atom of PdC_{60} (cf. Fig. 2.6).

Structures **1** and **2** both correspond to η^2 coordination of the two metal atoms over the same hexagon. In **2** both Pd atoms coordinate to (6–6′) junctions. This ensures an optimal bond for each metal atom to the fullerene and at the same time allows for some metal–metal interaction; r(Pd − Pd) = 3.01 Å. In **1**, on the other hand, one of the metal atoms coordinates to a (6–5) junction, and the bond dissociation energy of the second palladium atom drops by more than 10 kcal/mol compared to that of **2**, see Table 2.8. The reason for this may be sought in the occupancy of the donating lone-pair orbital for Pd. For the Pd atom adsorbed above the (6–6′) junction in structure **1** the number of electrons in this orbital (NBO population) is 1.69 (Table 2.8), only to increase to 1.76 if we consider the metal atom coordinated to a (6–5) junction. Clearly, the most effective back donation is achieved when the metal coordinates to (6–6′) sites. Structure **3** resembles that of **2**, except that the two metal atoms are further removed from each other. The BDE is only 2 kcal/mol lower than that of **2**, indicating the magnitude of the metal–metal interaction in the latter. Structures **4** and

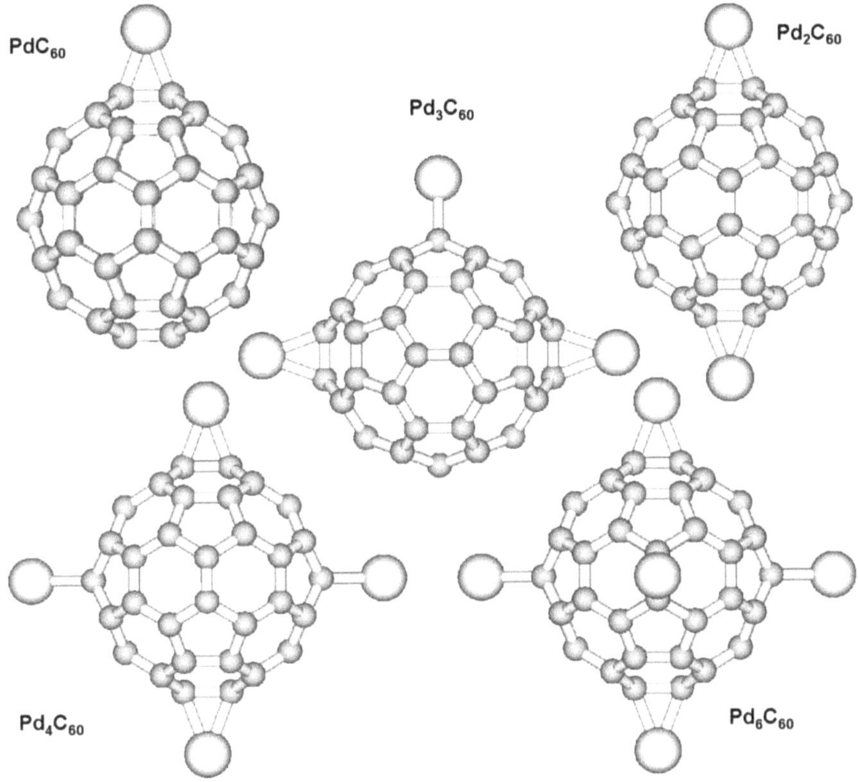

Fig. 2.6 Optimized exohedral metallofullerenes Pd_nC_{60} (n.1, 2, 3, 4, and 6)

Table 2.8 Bond dissociation energies (BDE) of PdC_{60}–Pd bonds and electronic properties of Pd_2C_{60}

Structure[a]	1	2	3	4	5
BDE[b](kcal/mol)	23.0(2.0)	34.6(1.9)	32.4(1.8)	10.0(0.5)	33.8(1.7)
Pd_1 $4d_\pi$ population[c]	1.69	1.75	1.63	1.61	1.69
Charge on Pd_1(e)[c]	0.29	0.26	0.29	0.17	0.22

5 describe coordination of a well-defined palladium dimer (Pd_2) in a singlet spin state, either end-on (**4**) or side-on (**5**), the latter structure featuring η^1 coordination mode to a (6–6′) junction of the fullerene molecule. The bond distances in the dimer are 2.76 and 2.63 Å in **4** and **5**, respectively. The end-on structure supports a very low BDE for the second palladium atom, reflecting that it binds to a less electronegative metal atom rather than directly to the fullerene. The two most stable structures found, (**2** and **5**), both maintain a Pd–Pd interaction while allowing both metal atoms to adsorb at (6–6′) junctions.

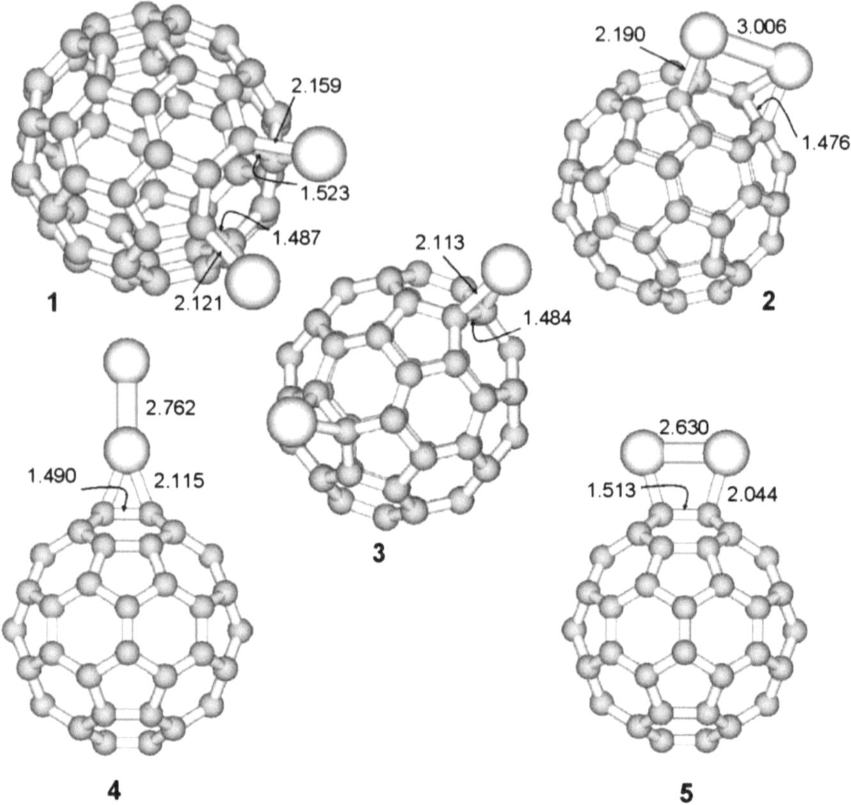

Fig. 2.7 Optimized exohedral metallofullerenes Pd_2C_{60}. *Curved* (*straight*) *arrows* point to C–C (C–Pd) bonds

For the Pd dimer, the ground state was identified as a triplet spin state, having an equilibrium bond length of 2.53 Å. The computed bond dissociation energy (20.5 kcal/mol) is in qualitative agreement with the experimental values (17 $\pm$ 4, 26 $\pm$ 5) [69]. Furthermore, it is much lower than twice the Pd-fullerene bond energy and clearly demonstrates that metal dimerization can not compete with metal adsorption onto the fullerene.

2.2.4 Conclusions

Density functional theory calculations show that the Pd–C_{60} bond energy remains essentially constant for up to six palladium atoms that are individually adsorbed onto C_{60}. A novel $Pd_2(\eta^2\text{-}C_{60})$ structure (**2** in Fig. 2.7) has been identified as the most stable arrangement of two palladium atoms on the external surface of C_{60}. It shows the two metal atoms bridging over a single six-membered ring, with both atoms benefiting from η^2 coordination at (6–6′) junctions as well as some metal–metal interaction. However, the difference in enthalpy between **2** and competing structures without metal–metal interaction is only 2 kcal/mol. Entropy considerations suggest that both isolated atoms and weakly bonded metal aggregates may exist in equilibrium. Binding of Pd atoms to the fullerene is preferred over palladium dimerization.

References

1. Sulman, E., Matveeva, V., Semagina, N., Yanov, I., Bashilov, V., Sokolov, V.: J. Mol. Catal. **146**, 257 (1999)
2. Claridge, J., Douthwaite, R., Green, M.: J. Mol. Catal. **89**, 113 (1994)
3. Nagashima, H., Nakaoka, A., Tajima, S., Saito, Y., Itoh, K.: Chem. Lett. **7**, 1361 (1992)
4. Nagashima, H., Kato, Y., Yamaguchi, H., Kimura, E., Kawanishi, T., Kato, M.: Saito, Y., Haga, M., Itoh, K. Chem. Lett. **7**, 1207 (1994)
5. Wohlers, M., Herzog, B., Belz, T., Bauer, A., Braun, T., Ruhle, T., Schlogl, R.: Synth. Met. **77**, 55 (1996)
6. Braun, T., Wohlers, M., Belz, T., Nowitzke, G., Wormann, G., Uchida, Y.: Pfander, N., Schlogl, R. Catal. Lett. **43**, 167 (1997)
7. Fagan, P., Calabrese, J., Malone, B.: Science **252**, 1160 (1991)
8. Fagan, P., Calabrese, J., Malone, B.: Acc. Chem. Res. **25**, 134 (1992)
9. Bashilov, V., Petrovskii, P., Sokolov, V., Lindeman, S., Guzey, I., Struchkov, Y.: Organometallics **12**, 991 (1993)
10. Dresselhaus, M., Dresselhaus, G., Eklund, P.: Science of Fullerenes and Carbon Nanotubes. Academic press Inc., San Diego (1996)
11. Lerke, S.A., Evans, D.H., Fagan, P.J.: J. Am. Chem. Soc. **114**, 7807 (1992)
12. Fagan, P.J., Calabrese, J.C., Malone, B.: J. Am. Chem. Soc. **113**, 9408 (1991)
13. Talyzin, A.V., Jansson, U.: Thin Solid Films **429**, 96 (2003)
14. Fujimoto, H., Nakao, Y., Fukui, K.: J. Mol. Struct. **300**, 425 (1993)
15. Lichtenberger, D., Wright, L., Gruhn, N., Rempe, M.: J. Organomet. Chem. **478**, 213 (1994)
16. Andriotis, A., Menon, M.: Phys. Rev. **60**, 4521 (1999)
17. Alemany, M., Dieguez, O., Rey, C., Gallego, L.: J. Chem. Phys. **114**, 9371 (2001)
18. Handy, N., Cohen, A.: J. Mol. Phys. **99**, 403 (2001)
19. Dunning Jr, T.H.: J. Chem. Phys. **53**, 2823 (1970)
20. Dunning, T., Hay, P.: Methods of Electronic Structure Theory. In: Schaeffer, H. (ed.) Modern theoretical chemistry. Plenum Press, New York (1977)
21. Magnusson, E., Schaefer III, H.: J. Chem. Phys **83**, 5721 (1985)
22. Dolg, M., Wedig, U., Stoll, H., Preuss, H.: J. Chem. Phys. **86**, 866 (1987)
23. Andrae, D., Haeussermann, U., Dolg, M., Stoll, H.: Theor. Chim. Acta **77**, 123 (1990)
24. Becke, A.: J. Chem. Phys. **98**, 5648 (1993)
25. Boys, S., Bernardi, F.: Mol. Phys. **19**, 553 (1970)
26. Koch, W., Holthausen, M.: A Chemist's Guide to Density Functional Theory. Wiley-VCH, Weinheim (2000)
27. Hay, P.: J. Chem. Phys. **66**, 4377 (1977)
28. Moore, C.: Atomic Energy Levels. NBS, Washington (1958)
29. Glendening, E., Reed, A., Carpenter, J., Weinhold, F.: NBO Version 3.1. (2001)
30. Dapprich, S., Frenking, G.: J. Phys. Chem. **99**, 9352 (1995)
31. Gorelsky, S.I.: AOMix: Program for Molecular Orbital Analysis. York University, Toronto (1997). http://www.sg-chem.net/

32. Gorelsky, S., Lever, A.: J. Organomet. Chem **635**, 187 (2001)
33. Hedberg, K., Hedberg, L., Bethune, D., Brown, C., Dorn, H., Johnson, R., de Vries, M.: Science **254**, 410 (1991)
34. Nunzi, F., Sgamellotti, A., Re, N., Floriani, C.: Organometallics **19**, 1628 (2000)
35. Kameno, Y., Ikeda, A., Nakao, Y., Sato, H., Sakaki, S.: J. Phys. Chem. **109**, 8055 (2005)
36. Campanera, J., Munoz, J., Vazquez, J., Bo, C., Poblet, J.: Inorg. Chem. **43**, 6815 (2004)
37. Li, J., Schreckenbach, G., Ziegler, T.: Inorg. Chem. **34**, 3245 (1995)
38. Hawkins, J., Meyer, A., Lewis, T., Loren, S., Hollander, F.: Science **252**, 312 (1991)
39. Fagan, P., Ward, M., Calabrese, J.: J. Am. Chem. Soc. **111**, 1719 (1989)
40. Haser, M., Almlof, J., Scuseria, G.: Chem. Phys. Lett. **181**, 497 (1991)
41. Morokuma, K., Borden, W.: J. Am. Chem. Soc. **113**, 1912 (1991)
42. Blomberg, M., Siegbahn, P., Svensson, M.: J. Phys. Chem. **96**, 9794 (1992)
43. Gates, B.: Catalytic Chemistry. Wiley, New York (1992)
44. Blomberg, M., Siegbahn, P., Nagashima, U., Wennerberg, J.: J. Am. Chem. Soc. **113**, 424 (1991)
45. Visscher, L., Dyall, K.: Atomic Data Nucl. Data Tables **67**, 207 (1997)
46. Weisshaar, J.: ACS Symp. Ser. **530**, 208 (1993)
47. Ritter, D., Weisshaar, J.: J. Am. Chem. Soc. **112**, 6425 (1990)
48. Widmark, P., Roos, B., Siegbahn, P.: J. Phys. Chem. **89**, 2180 (1985)
49. Siegbahn, P., Brandemark, U.: Theor. Chim. Acta **69**, 119 (1986)
50. Chatt, J., Duncanson, L.: J. Chem. Soc. **1953**, 2939 (1953)
51. Dewar, M.: Bull. Soc. Chim. Fr. p. 71(1951)
52. Massera, C., Frenking, G.: Organometallics **22**, 2758 (2003)
53. Cheng, P., Nyburg, S.: Can. J. Chem. **50**, 912 (1972)
54. Dresselhaus, M.S., Dresselhaus, G., Eklund, P.C.: Science of Fullerenes and Carbon Nanotubes. Academic Press, San Diego (1996)
55. Nagashima, H., Nakaoka, A., Saito, Y., Kato, M., Kawanishi, T., Itoh, K.: The first organometallic polymer of buckminsterfullerene. J. Chem. Soc. Chem. Commun. **69**, 377–379 (1992)
56. Nagashima, H., Kato, Y., Yamaguchi, H., Kimura, E., Kawanishi, T., Kato, M.: Haga, M., Itoh, K.: Synthesis and reactions of organoplatinum compounds of C_{60}, $C_{60}Pt_n$. Chem. Lett. **7**, 1207–1210 (1994)
57. Ivanova, V.N.: Fullerene compounds with transition metals M_nC_{60}: preparation, structure, and properties. J. Struct. Chem. **41**, 135–148 (2000)
58. Wohlers, M., Herzog, B., Belz, T., Bauer, A., Braun, Th, Ruhle, Th, Schlogl, R.: Ruthenium-C_{60} compounds: Properties and catalytic potential. Synth. Met. **77**, 55–58 (1996)
59. Talyzin, A.V., Jansson, U.: A comparative Raman study of some transition metal fullerides. Thin Solid Films **429**, 96–101 (2003)
60. Alemany, M.M.G., Dieguez, O., Rey, C., Gallego, L.J.: A densityfunctional study of the structures and electronic properties of C_{59}Ni and C_{60}Ni clusters. J. Chem. Phys. **114**, 9371–9374 (2001)
61. Andriotis, A.N., Menon, M.: Geometry and bonding in small $(C_{60})_nNi_m$ clusters. Phys. Rev. B **60**, 4521–4524 (1999)
62. Nagashima, H., Yamaguchi, H., Kato, Y., Saito, Y., Haga, M., Itoh, K.: Facile cleavage of carbon-palladium bonds in $C_{60}Pd_n$ with phosphines and phosphites. An alternative route to $(\eta^2 - C_{60})PdL_2$ and discovery of fluxionarity suggesting the rotation of C_{60} on the PdL_2 species in solution. Chem. Lett. **12**, 2153–2156 (1993)
63. Nagashima, H., Nakaoka, A., Tajima, S., Saito, Y., Itoh, K.: Catalytic hydrogenation of olefins and acetylenes over $C_{60}Pd_n$. Chem. Lett. **377**, 1361–1364 (1992)
64. Andrae, D., Haeussermann, U., Dolg, M., Stoll, H., Preuss, H.: Energyadjusted ab initio pseudopotentials for the second and third row transition elements. Theor. Chim. Acta **77**, 123–141 (1990)
65. Sparta, M., Jensen, V.R., Borve, K.J.: Structure and stability of substitutional metallofullerenes of the first-row transition metals. Fuller. Nanotub. Carb. Nanostruct. **14**(2–3), 269–278 (2006)

66. Hedberg, K., Hedberg, L., Bethunde, D.S., Brown, C.A., Dorn, H.C., Johnson, R.D., Devries, M.: Bond lengths in free molecules of buckminsterfullerene, C_{60}, from gas-phase electron-diffraction. Science **254**, 410–412 (1991)
67. Chatt, J., Duncanson, L.A.: Olefin co-ordination compounds 3. Infrared spectra and structure-attempted preparation of acetylene complexes. J. Chem. Soc. **28**, 2939–2947 (1953)
68. Dewar, M.J.S.: A review of the π-complex theory. Bull. Soc. Chim. Fr. **18**, C71–79 (1951)
69. Lin, S.S., Strauss, B., Kant, A.: Dissociation energy of Pd_2. J. Chem. Phys. **51**, 2282–2283 (1969)

Chapter 3
Nonlinear Optical Properties of Fullerene Derivatives

During recent years, a significant progress has been observed in the field of synthesis and characterization of photoactive materials [1–6]. In particular,tremendous effort have been invested in design of materials exhibiting high nonlinear optical response. A plethora of organic [7, 8] and organometallic [9, 10] systems have been studied in this context. One of the most common routes to design of new molecules with high values of first-order hyperpolarizability (β) has its roots in so called two-state model proposed in late seventies [11]. Within this model, β is expressed in terms of dipole moment difference, transition intensity and energy difference between excited and ground electronic state. Various molecular systems of donor–acceptor (DA) type according to two-level model have been proposed in order to maximize β. However, it has been reported recently that A–A systems containing [60]fullerene may also exhibit high β values [12].

[60]Fullerene is an interesting electron acceptor in which efficient charge separation and slow charge recombination occur, when combined with photo- and/or electro-active moieties, due to its small reorganization energy. A plethora of [60] fullerene-based hybrid materials with porphyrins [13–19] tetrathiafulvalenes [20–27] and ferrocenes [28–31] have been synthesized and studied in the past. Carbazole is a known organic electron donor which exhibits photophysical properties such as photoconductivity as observed in poly-vinyl-carbazole [32–35]. On the other hand, 4,7-diphenyl-2,1,3-benzothiazole-based compounds are strong fluorescent dyes [36–39], in which the absorption and emission profiles can be easily tuned upon substitution of the aryl unit incorporated in the benzothiazole moiety. Additionally, triphenylamine (TPA)-based materials are widely known as excellent hole-transporters and electroluminescent components [40–46], while films of TPA have been used in organic-light-emitting-diodes (OLEDs) [47]. It is the goal of the present contribution to analyze the properties of dyads **1–3** (Fig. 3.1) as useful photonic materials in which [60] fullerene is combined with a carbazole, a benzothiazole and benzothiazole-triphenylamine moieties, respectively. The investigated systems have been recently synthesized [45, 46]. In all three materials, a pyrrolidine unit is incorporated onto the skeleton of [60] fullerene via the well-established 1,3-dipolar

O. Loboda, *Quantum-Chemical Studies on Porphyrins, Fullerenes and Carbon Nanostructures*, Carbon Nanostructures, DOI: 10.1007/978-3-642-31845-0_3,

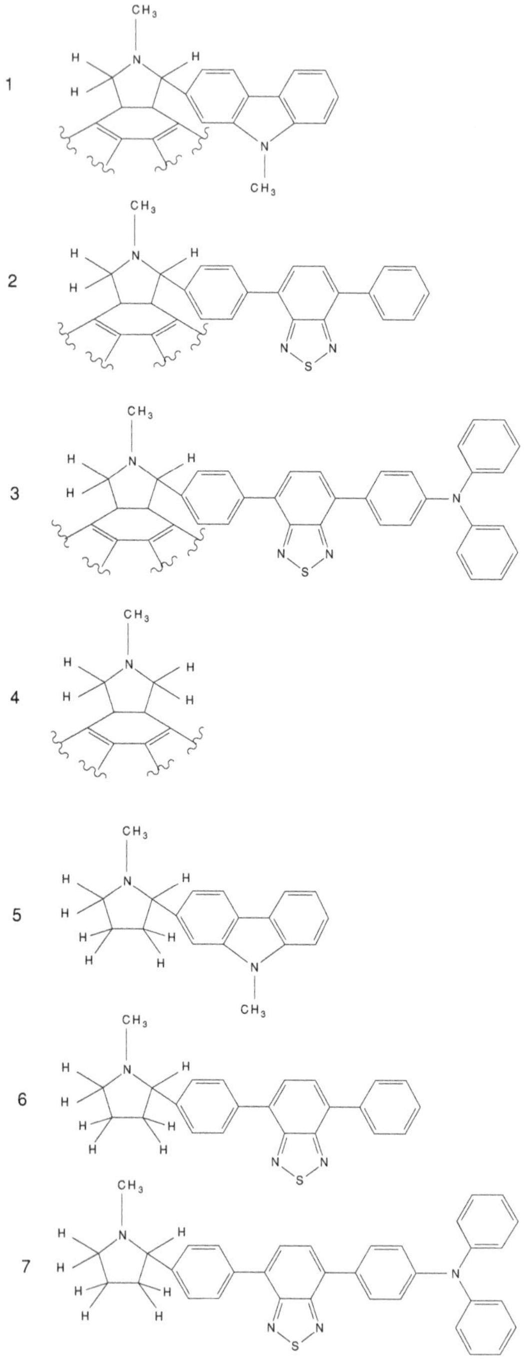

Fig. 3.1 Schematic representation of investigated molecules

cycloaddition reaction of azomethine ylides [48]. In particular, it is very interesting to study how modification of the chromophore influences the first- and the second-order hyperpolarizability as well as the two-photon absorption cross section of the whole dyad.

The performance of various theoretical approaches to calculations of nonlinear optical properties of molecules has been the subject of numerous studies [28–31]. It is well recognized nowadays that coupled-cluster hierarchy of methods gives very accurate values of molecular hyperpolarizabilities [50]. Undoubtedly, this level of theory is not the practical choice for computations for large systems. MP2 method is also computationally expensive, making it rather more suitable for benchmarking than for systematic studies of nonlinear optical properties of extended organic systems. The alternative is the density functional theory, which accounts for electron correlation and is much less computationally demanding than post Hartree–Fock electron correlation treatments. It has been recognized, however, that the majority of proposed exchange-correlation functionals failed to correctly predict polarizabilities and hyperpolarizabilities of conjugated chains [51, 52]. One of the strategies proposed recently to eliminate the deficiencies of existing functionals is namely, density functional theory accounting for long-range effects (further denoted as LR-DFT). LR-DFT has been used for calculations of nonlinear optical properties for several systems. However, we are not aware of any studies regarding calculations of hyperpolarizabilities of organofullerenes using LR-DFT. It is also the goal of the present contribution to assess the performance of various theoretical approaches for the calculation of nonlinear optical properties of [60]fullerene-chromophore dyads, including recently proposed functionals belonging to the LR-DFT group.

Two main aspects of the present contribution can be generalized and formulated as follows. Firstly, we compute both electronic (Sect. 3.2.2) and vibrational (Sect. 3.2.4) contributions to (hyper)polarizability. Thus, we explore the limitations of the currently available computational procedures. Secondly, we associate the linear and nonlinear optical properties of investigated organofullerenes with their electronic structure (Sect. 3.2.3). The purpose of the analysis of the relations between NLO properties and structure of organofullerenes is to make a basis for further rational design of new [60]fullerene derivatives suitable for photonic applications.

3.1 Computational Methods

3.1.1 Basic Definitions of L & NLO Properties

In the presence of static uniform electric field, the total energy of molecular system can be expressed as a Taylor series:

$$E(F) = E(0) - \mu_i F_i - \frac{1}{2!}\alpha_{ij} F_i F_j - \frac{1}{3!}\beta_{ijk} F_i F_j F_k - \frac{1}{4!}\gamma_{ijkl} F_i F_j F_k F_l - \cdots \quad (3.1)$$

where $E(0)$ denotes energy of a molecule without external perturbation. The Greek symbols $\alpha, \beta \ldots$, label tensor quantities. The expansion (3.1) is also known as the so-called T-convention [53].

The dipole moment of a molecule in the presence of uniform electric field is given by:

$$\mu_i(F) = \mu_i(0) + \alpha_{ij} F_j + \frac{1}{2!}\beta_{ijk} F_j F_k + \frac{1}{3!}\gamma_{ijkl} F_j F_k F_l + \cdots \tag{3.2}$$

Both Eqs. (3.1) and (3.2) can be used for determination of polarizability (α), first- (β) and second-order hyperpolarizability (γ). The averaged (hyper)polarizabilities used throughout the book are defined in the following way [28–31]:

$$\overline{\alpha} = \frac{1}{3} \sum_{i=x,y,z} \alpha_{ii}, \tag{3.3}$$

$$\overline{\beta} = \sum_{i=x,y,z} \frac{\mu_i \beta_i}{|\mu|}, \tag{3.4}$$

where

$$\beta_i = \frac{1}{5} \sum_{j=x,y,z} \left(\beta_{ijj} + \beta_{jij} + \beta_{jji}\right), \tag{3.5}$$

and

$$\overline{\gamma} = \frac{1}{15} \sum_{ij=x,y,z} (\gamma_{iijj} + \gamma_{ijji}) \tag{3.6}$$

Within the Born–Oppenheimer approximation, the (hyper)polarizabilities can be divided into two parts, namely electronic and vibrational [28–31]:

$$P = P^e + P^{vib}. \tag{3.7}$$

In this work we compute the nuclear relaxation contributions to α^{vib} and β^{vib}.

3.1.2 Electronic Contributions

The numerical differentiation of the total energy of a system (Eq. (3.1)) with respect to electric field allows for determination of expansion coefficients, i.e. α, β and γ [54]:

$$\mu_i F_i = -\frac{2}{3}[E(F_i - E(-F_i)] + \frac{1}{12}[E(2F_i) - E(-2F_i)], \tag{3.8}$$

$$\alpha_{ii} F_\alpha^2 = \frac{5}{2} E(0) - \frac{4}{3}[E(F_i) + E(-F_i)] + \frac{1}{12}[E(2F_i) + E(-2F_i)], \quad (3.9)$$

$$\beta_{iii} F_i^3 = [E(F_i) - E(-F_i)] - \frac{1}{2}[E(2F_i) + E(-2F_i)], \quad (3.10)$$

$$\gamma_{iiii} F_i^4 = -6E(0) + 4[E(F_i) + E(-F_i)] - [E(2F_i) + E(-2F_i)]. \quad (3.11)$$

It is also possible to use expansion (3.2) for calculation of α, β and γ. Nevertheless, in a great majority of calculations presented herein, we used expansion (3.1). Moreover, in few cases, diagonal component of hyperpolarizability β_{iii} was also calculated as a derivative of α_{ii} with respect to electric field. Whenever computationally possible, we used the Romberg procedure to remove the higher-order contaminations to β and γ [55].

The calculations of electronic contributions to (hyper)polarizabilities using Hartree–Fock and second-order Møller–Plesset perturbation theory (MP2) were performed using Gaussian 03 suite of programs [56]. Time-dependent Hartree–Fock (TDHF) method was employed at the semiempirical level of theory for calculations of α, β and γ as implemented in MOPAC 2007 package [46]. Density functional theory computations were performed using Gaussian 03 package (B3LYP, PW91, LDA) [56], ADF (CurDFT, LB94, RevPBEx, GRAC, KT1, KT2, KLI) [58–60], Dalton (CAM-B3LYP) [61] and GAMESS US (LC-BLYP) [62].

3.1.3 Vibrational Contributions

In order to compute the nuclear relaxation (NR) contribution, the field induced coordinates (FICs) approach has been employed [63, 64]. Within this methodology, these coordinates correspond to displacements of the normal coordinates which are associated with the equilibrium geometry in the presence of a static electric field. Applying the stationary condition to the potential energy surface, one may solve the resulting equation order by order for the i-th value of the field-free normal coordinate at the field-relaxed geometry. This allows to calculate the field-dependent normal coordinate given by the following formula [63]:

$$Q_i^F(F_x, F_y, F_z) = -\sum_a^{x,y,z} q_1^{i,a} F_a - \sum_{a,b}^{x,y,z} \left[q_2^{i,ab} - \sum_{j=1}^{3N-6} \frac{a_{21}^{ij,a}}{a_{20}^{ii}} q_1^{i,b} + \sum_{j,k=1}^{3N-6} \frac{3a_{30}^{ijk}}{2a_{20}^{ii}} q_1^{j,a} q_1^{k,b} \right] F_a F_b \quad (3.12)$$

The sum runs over $3N$-6 normal coordinates. The order of the displacements in the presence of the field determines the corresponding order of the FIC. Depending on the NLO property of interest, a number of FICs is used in order to calculate the NR contribution, which is equivalent to the value obtained during $3N$-6 normal coordinate calculations. Using the first-order FIC one can compute the static nuclear relaxation contributions to the polarizability and first hyperpolarizability, as given by the following analytical formulae:

$$\alpha_{ab}^{nr}(0;0) = \frac{1}{2}\sum_{i=1} P_{ab} a_{11}^{i,a} q_1^{i,b} = [\mu^2]^{(0,0)} \tag{3.13}$$

$$\begin{aligned}\beta_{abc}^{nr}(0;0,0) &= \sum_{i=1} P_{abc} a_{12}^{i,ab} q_1^{i,b} - \sum_{i,j=1} P_{abc} a_{21}^{ij,a} q_1^{i,b} q_1^{j,c} \\ &\quad - \sum_{i,j,k=1} P_{abc} a_{30}^{ijk} q_1^{i,a} q_1^{j,b} q_1^{k,c} \\ &= [\mu\alpha]^{(0,0)} + [\mu^3]^{(1,0)} + [\mu^3]^{(0,1)}\end{aligned} \tag{3.14}$$

where $P_{ab...}$ indicates the permutation over indices $a, b, \ldots, q_1$ is the first-order FIC, a_{nm} involve derivatives of the potential energy V(Q,F) with respect to the n-th normal coordinate and derivatives with respect to the m-th field component and in the notation $[A]^{(l,k)}$, the superscripts l and k denote the level of the electrical and mechanical anharmonicities, respectively [65]. Likewise, it is possible to compute frequency dependent NR contribution for the Pockels effect $\beta(-\omega;\omega,0)_{\omega\to\infty}$ and electric field induced second-harmonic generation (EFISH) $\gamma(-2\omega;\omega,\omega,0)$ using first-order FIC. More details about the FICs method can be found elsewhere [63, 64]. All the necessary energy and property derivatives were computed numerically at the optimum geometry, by applying a number of steps along the FIC vector. The calculations were performed using the Gaussian 03 program[56]. Due to the computational limitations, the vibrational contributions to α and β for **1**, **2** and **3** were estimated based on the double-harmonic oscillator approximation [28–31]:

$$[\mu^2]_{ij}^{(0,0)} = \sum_{\xi}\left(\frac{1}{\omega_\xi^2}\right)\left[\left(\frac{\partial \mu_i}{\partial Q_\xi}\right)_0\left(\frac{\partial \mu_j}{\partial Q_\xi}\right)_0\right], \tag{3.15}$$

$$\begin{aligned}[\mu\alpha]_{ijk}^{(0,0)} &= \sum_{\xi}\left(\frac{1}{\omega_\xi^2}\right)\left[\left(\frac{\partial \mu_i}{\partial Q_\xi}\right)_0\left(\frac{\partial \alpha_{jk}}{\partial Q_\xi}\right)_0\right. \\ &\quad \left.+\left(\frac{\partial \mu_j}{\partial Q_\xi}\right)_0\left(\frac{\partial \alpha_{ik}}{\partial Q_\xi}\right)_0+\left(\frac{\partial \mu_k}{\partial Q_\xi}\right)_0\left(\frac{\partial \alpha_{ij}}{\partial Q_\xi}\right)_0\right],\end{aligned} \tag{3.16}$$

where Q_ξ denotes ξ-th normal coordinate and summation runs over all normal modes of vibration. The calculations of $[\mu^2]^{(0,0)}$ and $[\mu\alpha]^{(0,0)}$ terms were performed using the GAMESS US package [62]. One should remember that the calculations of vibrational contributions at a given level of theory require the use of geometry optimized using the very same method.

3.1.4 Two-Photon Absorption

The quantity which describes a process of simultaneous absorption of two-photons of different energy ($\hbar\omega_1 \neq \hbar\omega_2$) with different polarization ($\imath_1 \neq \imath_2$) at the molecular level is given by the following Eq. [66]:

$$S_{ij}^{0F} = \hbar^{-1} \sum_K \left[\frac{\langle 0|\bar{\zeta}_1 \hat{\mu}_i|K\rangle\langle K|\bar{\zeta}_2 \hat{\mu}_j|F\rangle}{\omega_K - \omega_1} + \frac{\langle 0|\bar{\zeta}_2 \hat{\mu}_i|K\rangle\langle K|\bar{\zeta}_1 \hat{\mu}_j|F\rangle}{\omega_K - \omega_2} \right] \tag{3.17}$$

where $\hbar\omega_1 + \hbar\omega_2$ should satisfy the resonance condition and $\langle K|\imath_1 \hat{\mu}|L\rangle$ is the transition moment between electronic states K and L, respectively.

Since in most experiments one source of photons is used, one can substitute the angular frequencies ω_1 and ω_2 for $0.5 \cdot \omega_F$. In the case of isotropic media the averaged two-photon absorption cross section is given by:

$$\begin{aligned} \langle \delta^{0F} \rangle &= \langle |S_{ij}^{0F}(\bar{\zeta}_1, \bar{\zeta}_2)|^2 \rangle \\ &= \frac{1}{30} \sum_{ij} \left[S_{ii}^{0F} (S_{jj}^{0F})^* F + S_{ij}^{0F} (S_{ij}^{0F})^* G + S_{ij}^{0F} (S_{ij}^{0F})^* H \right] \end{aligned} \tag{3.18}$$

where $F = F(\bar{\zeta}_1, \bar{\zeta}_2)$, $G = G(\bar{\zeta}_1, \bar{\zeta}_2)$, $H = H(\bar{\zeta}_1, \bar{\zeta}_2)$ are the polarization variables. For two linearly polarized photons with parallel polarization, $F, G, H = 2$ and the two-photon absorption cross-section is [66]:

$$\langle \delta^{0F} \rangle = \frac{1}{15} \sum_{ij} \left[S_{ii}^{0F} (S_{jj}^{0F})^* + 2 S_{ij}^{0F} (S_{ij}^{0F})^* \right]. \tag{3.19}$$

The geometry optimizations of molecular structures were performed using the semi-empirical PM3 Hamiltonian [67, 68] and the B3LYP/6-31G(d) method. All stationary points were confirmed to be true minima by evaluation of hessian. In order to speed up HF and DFT calculations, the fast multipole method (FMM) [47, 48, 69] has been used as implemented in Gaussian suite of programs [56]. We also used linear scaling approaches for calculations of nonlinear optical properties as implemented in ADF package [58]. All the properties are expressed in atomic units. Conversion factors can be found elsewhere [28–31].

3.2 Fullerene-Benzothiadiazole and -Carbazole Derivatives

This section is organized as follows. Firstly, we shall discuss the structure of the investigated systems (Sect. 3.2.1). The electronic contributions to polarizability, first- and second-order hyperpolarizability are presented in Sect. 3.2.2. In the next section we analyze one- and two-photon absorption spectra of investigated molecules (Sect. 3.2.3), while the discussion of vibrational contributions is given in Sect. 3.2.4.

3.2.1 Structure of Investigated Molecules

Due to the size of the investigated systems, the calculation of hessian is computationally demanding even using density functional theory. Thus, we used the PM3 method for geometry optimizations. In order to assess the reliability of structures optimized using the PM3 method, we optimized **1** (Fig. 3.2) at the B3LYP/6-31G(d) level of theory as well. The comparison of PM3 and B3LYP results reveals only very minor changes in the bond distances of fullerene moiety, i.e. carbon–carbon (6–5) and (6–6) bond length differences for both methods do not exceed 0.01 and 0.012 Å, respectively. The values of selected geometrical parameters for **1** are listed in Table 3.1. Both methods give very similar results for the N–CH_3 distance and for the C–C bond lengths in pyrrolidine ring. The most significant differences occur in the carbazole moiety. The results of PM3 and B3LYP calculations show that the N'–CH_3 group of carbazole is closer to the plane of pyrrolidine ring for the latter method: The corresponding dihedral angle for both methods is 149.4 and 177.8 °, respectively. Since no experimental structure is available for **1** the data presented in Table 3.1 illustrate the level of uncertainty for both theoretically determined molecular geometries. Moreover, calculations were also performed for unsubstituted [60]fullerene molecule. The comparison of the bond-lengths calculated using PM3 method with the electron diffraction data [72] reveals no significant differences between theory (1.458 and 1.384 Å) and experiment (1.458 and 1.401 Å). Recent DFT calculations provide the values of bond-lengths 1.465 and 1.411 Å[73].

In Table 3.2 we present the values of dipole moments, first-, second- and third-order polarizability of **1** calculated at the HF/6-31G(d) level of theory. The properties were evaluated for two structures, which were optimized using PM3 and B3LYP/6-31G(d) methods. The influence of the level of theory, at which the geometry of **1** was optimized, on the values of $\overline{\alpha}$, $\overline{\beta}$ and $\overline{\gamma}$ is not significant. In the case of all computed properties, however, their values for two different geometries do not differ substantially. The structures of all investigated molecules were optimized by using the PM3 method, because the effect of the geometries on the linear and nonlinear optical properties in not significant and the computational cost of DFT optimizations is rather large. It should be mentioned that we consistently use the so-called standard orientation for studied systems [74].

3.2.2 Electronic Contributions to (Hyper)Polarizabilities

In this section, we shall discuss the electronic contributions to α, β and γ. In particular, two following aspects will be considered in detail: the assessment of reliability of results obtained using various exchange-correlation functionals and the relation between the structure of investigated systems and their non-linear optical properties. Before we proceed to analysis of NLO properties of investigated systems, it is inter-

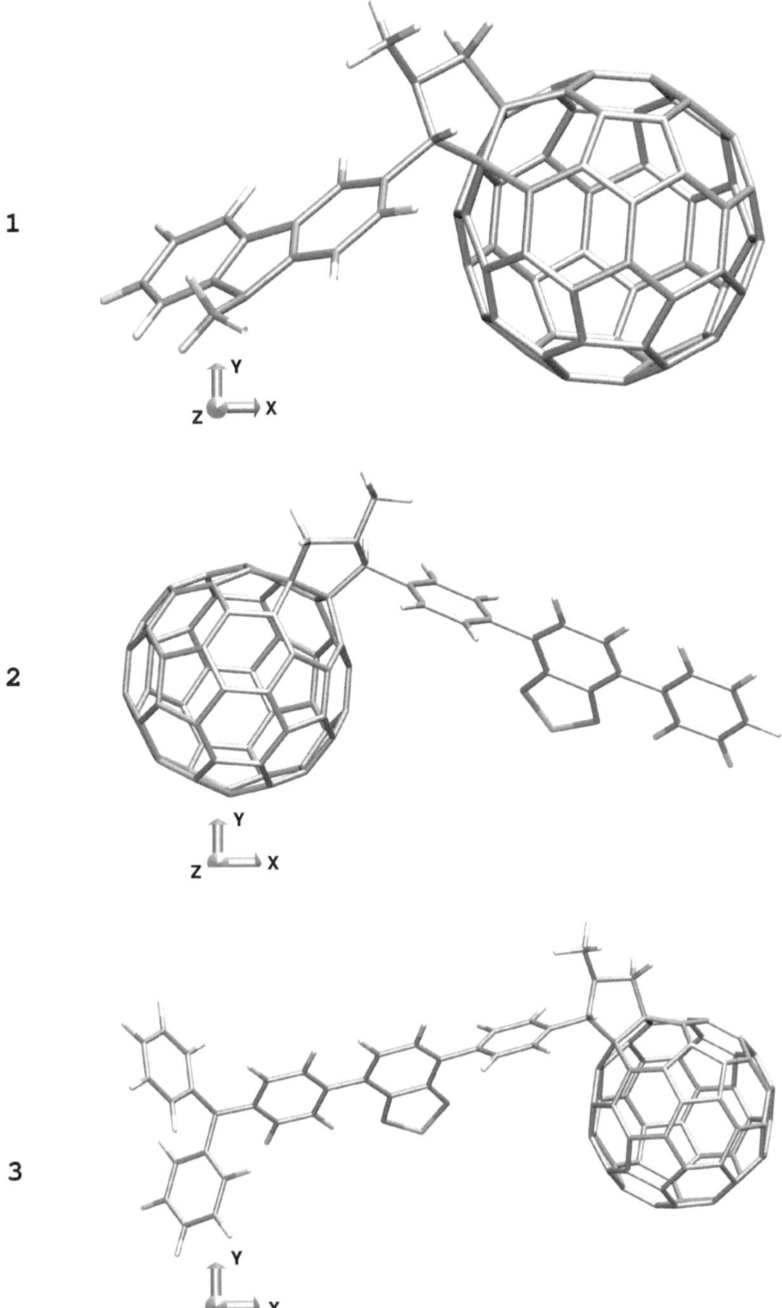

Fig. 3.2 Structure of investigated [60]fullerene derivatives. Shown is the orientation in cartesian coordinates used in property calculations

Table 3.1 Geometrical parameters of **1** calculated using PM3 method and at the B3LYP/6-31G(d) level of theory

Method	C–C(6–6)	C–CH_2	C–CHR	N–CH_2	N–CHR	N–CH_3	N'–CH_3	∠CH_2NCH_3	∠CN'CH_3
PM3	1.584	1.537	1.556	1.486	1.500	1.474	1.466	113.8	121.0
B3LYP	1.611	1.556	1.558	1.452	1.463	1.457	1.447	112.7	125.6

N' refers to nitrogen atom located in carbazole moiety. Distances are given in Åand angles in degrees

Table 3.2 Polarizability α, first hyperpolarizability β, second hyperpolarizability γ and dipole moment μ of **1** calculated at the HF/6–31G(d) level of theory for the structures optimized using PM3 and B3LYP/6–31G(d) methods

	PM3 geometry	B3LYP/6–31G(d) geometry
μ	1.72	1.92
$\overline{\alpha}$	634.09	640.33
$\overline{\beta}$	403.33	470.14
$\overline{\gamma}$	71279	70414

All values are given in atomic units

esting to check the performance of various theoretical methods in calculations of α and γ for unsubstituted [60]fullerene molecule.

In Table 3.3, the calculated and experimental values of the static polarizability and second hyperpolarizability of [60]fullerene are presented. The value of α_{xx} calculated using PM6 method is much closer to the experimental value than the PM3 value. The value of $\overline{\alpha}$ calculated at the MP2 level of theory seems to be slightly underestimated in comparison with the experimental data. So is the MCSCF value of α_{xx}. It is very interesting to note that results for LDA and GGA functionals fall in the range of experimental uncertainty. Density functional theory accounting for long-range effects gives significantly underestimated values of polarizability. In the case of second hyperpolarizability, both LR-DFT and semiempirical methods significantly underestimate its value. LDA and GGA functionals give slightly larger values in comparison with experimental data. Based on the results presented in Table 3.3, one may arrive at conclusion that MP2 calculations can be considered as reference point for larger systems for which neither advanced correlation treatments are feasible to perform nor experimental data exists.

The performance of various functionals commonly employed within density functional theory in prediction of electrooptic properties of **1** is presented in Table 3.4. It is interesting to check the effect of basis set extension on μ, α and β. At the HF and MP2 levels of theory, one observes that the absolute values of μ_x and β_{xxx} decrease while the value of α_{xx} increases upon basis set enlargement (3-21G vs 6-31G(d)). The influence of electron correlation on β_{xxx} is significant, i.e. its value at the MP2 level of theory is almost two times larger than the value calculated using HF method. Similar influence of electron correlation on β_{xxx} was observed for other organic molecules [75, 76]. The MP2 results shall serve as a reference point for assessment of DFT calculations of hyperpolarizability. It is now well recognized that traditional functionals commonly used in DFT suffer from what is known as shortsightedness

Table 3.3 Polarizability α and second hyperpolarizability γ of [60]fullerene molecule. $\alpha_{xx} = \alpha_{yy} = \alpha_{zz} = \overline{\alpha}$ and $\gamma_{xxxx} = \overline{\gamma}$

	PM3	PM6	CAM–B3LYP/3-21G	MP2/6-31G(d)	LDA[114]	MCSCF [115]	B3LYP[116]	LB94 [116]	PBE[116]	Expt
α_{xx}	431.39	527.62	422.81	462.52	571.58	506.80	547.0	544.10	554.90	579–595 [117–120] 516 $\pm$ 54 [121] 533 $\pm$ 27 [122]
γ_{xxxx}	64060	52795	23307	121632	137950	109200	118270	87020	119230	93000 $\pm$ 14000[123]

All values are given in atomic units

Table 3.4 Dipole moment μ_x, electronic first- (α_{xx}) and second-order polarizability (β_{xxx}) of **1**

Method	$\mu/\alpha/\beta^{\ddagger}$	μ_x	α_{xx}	β_{xxx}
HF/3–21G	$\mathcal{A}/\mathcal{A}/\mathcal{A}$	−1.79	681.87	−661.1
HF/6–31G(d)	$\mathcal{A}/\mathcal{E}/\mathcal{E}$	−1.47	735.27	−558.1
MP2/6–31G(d)	$\mathcal{E}/\mathcal{E}/\mathcal{E}$	−1.67	782.99	−1301.2
MP2/3–21G	$\mathcal{E}/\mathcal{E}/\mathcal{E}$	−1.88	719.14	−1525.6
CAM-B3LYP/3–21G	$\mathcal{A}/\mathcal{A}/\mathcal{A}$	−1.82	719.27	−1671.9
LC-BLYP/3–21G	$\mathcal{E}/\mathcal{E}/\mathcal{E}$	−1.77	702.79	−1193.4
B3LYP/3–21G	$\mathcal{A}/\mathcal{A}/\mathcal{E}$	−1.91	758.63	−4484.7
CurDFT/DZP	$\mathcal{A}/\mathcal{A}/\alpha$	−2.09	672.05	−5378.5
LB94/DZP	$\mathcal{A}/\mathcal{A}/\mathcal{A}$	−2.09	901.25	−10291.0
RevPBEx/DZP	$\mathcal{A}/\mathcal{A}/\mathcal{A}$	−2.42	900.59	−15196.0
GRAC/DZP	$\mathcal{A}/\mathcal{A}/\mathcal{A}$	−2.64	926.19	−19816.0
PW91/3–21G	$\mathcal{A}/\mathcal{A}/\mathcal{E}$	−2.19	819.73	−20125.6
KT2/DZP	$\mathcal{A}/\mathcal{A}/\mathcal{A}$	−2.79	939.00	−22050.0
LDA/3–21G	$\mathcal{A}/\mathcal{A}/\alpha$	−2.32	830.09	−22487.0
KT1/DZP	$\mathcal{A}/\mathcal{A}/\mathcal{A}$	−2.81	948.14	−22541.0
KLI/DZP	$\mathcal{A}/\mathcal{A}/\alpha$	−3.65	975.11	−83807.0

All values are given in atomic units

‡ $\mathcal{A}\alpha\mathcal{E}$ stand for fully analytical calculation, calculation as a derivative of the polarizability and calculation as a derivative of the energy, respectively

[77]. The incorrect description of charge-transfer excited states [9], the unrealistic evolution of hyperpolarizabilities with increasing size of π-delocalized molecules [51, 52] and the wrong prediction of bond length alternation for conjugated systems [79], all these drawbacks are related to spurious interaction of an electron with itself [80]. Several promising routes have been proposed to circumvent the above mentioned deficiencies. In this context, the LR-DFT (including long-range corrected functionals [81–83] and Coulomb-attenuating theory [84]), current DFT [85] and KLI [86] approaches should be mentioned.

We are not aware of any systematic study of DFT performance in calculations of NLO properties of [60]fullerene-chromophore dyads. Hence, it is very interesting to compare the DFT values of hyperpolarizabilities with *ab initio* MP2 results. In Table 3.4 the values of μ_x, α_{xx} and β_{xxx} for **1** are presented. The DFT data are ordered according to decreasing accuracy of the latter quantity. As it is clearly seen, the accuracy of LR-DFT values of β_{xxx} outperforms the results obtained by remaining functionals. The CAM-B3LYP value of β_{xxx} is about 150 a.u. larger than MP2 value, while the difference between LC-BLYP and MP2 results is over two times larger. The values of β_{xxx} calculated using four approaches, namely B3LYP, SAOP, CurDFT and LB94, are few times grater than the MP2/3-21G reference value. All remaining functionals give β_{xxx} values that are order of magnitude larger than MP2 results.

In Table 3.5, the results of calculations of μ_x, α_{xx} and β_{xxx} for several model systems are presented. The systems have been selected according to the electron donating character of substitutes. Two functionals were selected for comparison, namely CAM-B3LYP and LB94. The former provides the value of β_{xxx} for **1** close

Table 3.5 The values of dipole moment (μ), first- (α), second-order polarizability (β) for several model systems

Label	Molecule	μ_x	α_{xx}	β_{xxx}	Method
8	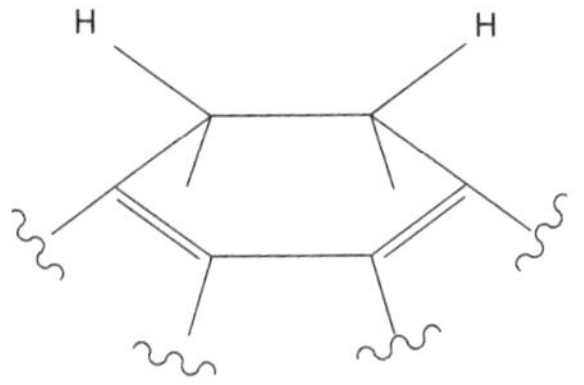 H, H	1.00	466.85	−99.22	HF/6–31G(d)
		1.15	526.63	−111.6	LB94/DZP
		1.17	449.30	−81.61	CAM-B3LYP/3–21G
9	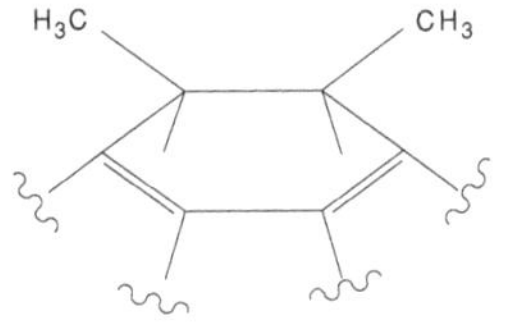 H_3C, CH_3	1.18	510.68	−31.1	HF/6–31G(d)
		1.56	589.18	−28.1	LB94/DZP
		1.47	494.28	80.9	CAM-B3LYP/3–21G
10	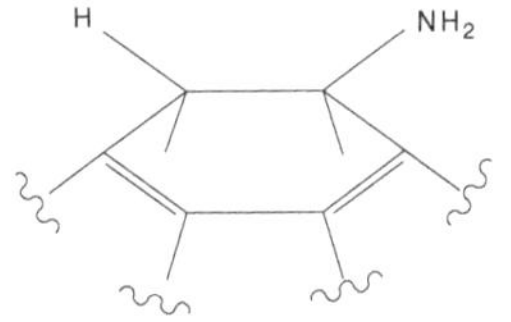 H, NH_2	0.90	484.45	−46.9	HF/6–31G(d)
		1.03	551.87	99.8	LB94/DZP
		1.14	466.84	73.1	CAM-B3LYP/3–21G
11	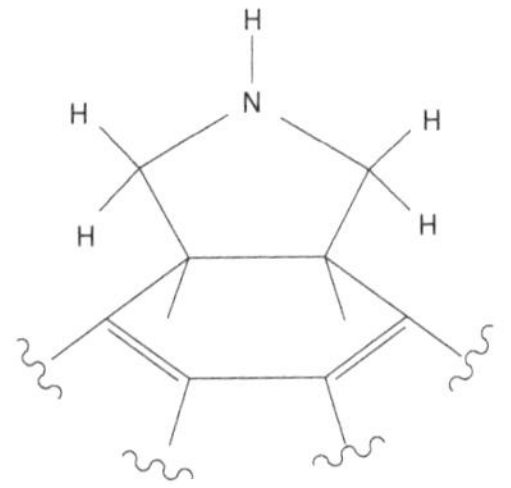 H, N, H, H, H, H	0.72	498.65	16.2	HF/6–31G(d)
		0.81	522.36	97.4	MP2/6–31G(d)
		1.13	581.40	2038.6	LB94/DZP
		0.93	481.30	279.19	CAM-B3LYP/3–21G
12	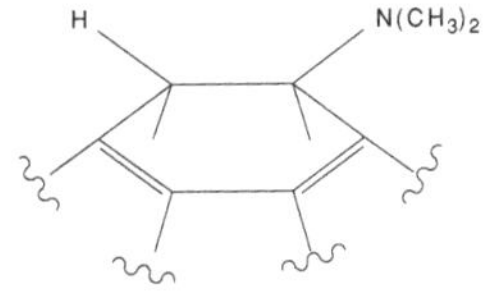 H, $N(CH_3)_2$	1.05	531.71	−128.95	HF/6–31G(d)
		1.42	622.81	2694.6	LB94/DZP
		1.28	516.54	123.64	CAM-B3LYP/3–21G

All values are given in atomic units

Table 3.6 The values of dipole moment (μ), first- (α), second- (β) and third-order polarizability (γ) for **1**, **2** and **3**

Property	Method	1	2	3
μ_x	HF/3-21G	−1.79	1.09	−1.29
	MP2/3-21G	−1.88	1.00	−1.25
	LC-BLYP/3-21G	−1.77	0.94	−1.22
	PM3	−1.30	0.86	−1.02
	PM6	−1.72	0.95	−1.18
α_{xx}	HF/3-21G	681.87	859.68	1051.74
	MP2/3-21G	719.14	867.34	1079.56
	LC-BLYP/3-21G	702.79	874.92	1091.88
	PM3	671.91	858.72	1061.88
	PM6	818.24	1743.86	1548.06
$\overline{\alpha}$	HF/3-21G	589.43	675.63	812.12
	PM3	591.89	679.7	817.8
β_{xxx}	HF/3-21G	−674.0	664.1	−4552.2
	MP2/3-21G	−1525.6	985.4	−8267.3
	LC-BLYP/3-21G	−1193.3	874.9	−7457.4
	PM3	−1028.2	1373.7	−7400.9
β_{yyy}	PM3	−59.5	163.5	440.9
β_{zzz}	PM3	196.9	−24.9	−102.6
$\overline{\beta}$	PM3	546.6	521.5	1746.2
γ_{xxxx}	HF/3-21G	171729	596479	1542361
	MP2/3-21G	384606	988307	3103560
	LC-BLYP/3-21G	291518	826348	2655793
	PM3	261749	1073875	3051099
$\overline{\gamma}$	PM3	135378	307891	730346

All values are given in atomic units

Table 3.7 The comparison of linear and nonlinear optical properties of **1**, **2**, **3** and **5**, **6** and **7**

	Method	**1**	**5**	**2**	**6**	**3**	**7**
$\overline{\alpha}$	PM3	591.89	166.48	679.70	250.77	817.79	385.46
$\overline{\beta}$	PM3	546.63	44.70	521.46	−50.58	1746.22	−388.94
$\overline{\gamma}$	PM3	135377	38376	307890	174300	730346	507853

All values are given in atomic units

to MP2 results, while the latter significantly overestimates the value of β_{xxx}. As it is seen, the large discrepancy between CAM-B3LYP and LB94 results is only observed for **11** and **12**. This behaviour may be certainly attributed to electron donating capabilities of substituents. In the case of [60]fulleropyrrolidine molecule, the value of β_{xxx} calculated using LB94 functionals is almost twenty times larger than the MP2 value. Based on the data presented in Tables 3.4 and 3.5, one may arrive at conclusion that LR-DFT is the only reasonable choice, among studied functionals, for reliable prediction of NLO properties for this type of systems.

The results of calculations of electrooptic properties for **1**, **2** and **3** are presented in Table 3.6. Due to the computational limitations, MP2 results are presented only for single component (x) of μ, α, β and γ. Since rotationally invariant hyperpolarizabilities are more suitable for structure-property relationships analysis, one have to select reliable level of theory for calculations of $\overline{\alpha}$, $\overline{\beta}$ and $\overline{\gamma}$. The values of α_{xx}, β_{xxx} and γ_{xxxx} calculated at the MP2 level of theory may serve as a reference data. Based on the results presented in Tables 3.3 and 3.6 the following ordering of α_{xx} values can be observed: $\alpha_{xx}^{\mathrm{PM6}} > \alpha_{xx}^{\mathrm{MP2}} > \alpha_{xx}^{\mathrm{PM3}}$. It is worth noticing, that PM6 model gives the values of α_{xx} significantly overestimated. On the other hand, the values of α_{xx}, β_{xxx} and γ_{xxxx} predicted by the PM3 model may be assumed as qualitatively correct. Thus, Table 3.6 involves mainly two pieces of information: Firstly, it demonstrates that PM3 method provide satisfactory (at least at a semiquantitative level) property values for μ_x, α_{xx}, β_{xxx} and γ_{xxxx} for **1**, **2** and **3**. Secondly, it contains the average values of α, β and γ of **1**, **2** and **3** calculated using PM3 methods. We shall use values of $\overline{\beta}$ and $\overline{\gamma}$ calculated using PM3 method as the basis for the further analysis. It follows from Table 3.6 that diagonal β_{xxx} component is much larger than the averaged hyperpolarizability, i.e. in the case of **3**, the absolute value of the ratio $\beta_{xxx}/\overline{\beta}$ is as large as 4.2. It clearly indicates the importance of other components of β tensor. It is very interesting to note, that the value of $\gamma_{xxxx}/\overline{\gamma}$ for **3** is identical as in the case of first-order hyperpolarizability. The comparison of averaged first-order hyperpolarizability for **1**, **2** and **3** shows that both **1** and **2** are characterized by similar values of $\overline{\beta}$. In both cases (**1** and **2**) [60]fullerene serves as the electron acceptor. However, it should be mentioned that benzothiadiazole group has also electron withdrawing capabilities [38]. The value of $\overline{\beta}$ for **3** is three times larger than that of **1** and **2**. This enhancement is certainly due to the larger electron donating capability of the chromophore. In the case of $\overline{\gamma}$, however, one observes systematic increase among the following sequence: **1**, **2** and **3**. It should be noted, that the β_{xxx} value of p-nitroaniline molecule, which is considered to be model charge-transfer system, calculated at the MP2/6-31+G(d) level of theory is similar to that of **1** [76]. The comparison of experimental value of $\overline{\gamma}$ for [60]fullerene (see Table 3.3) with that of **1** reveal, that the difference does not exceed 30 %. The averaged values of α, β and γ for the [60]fullerene-chromophore dyads and for the chromophores are presented in Table 3.7. It is clearly seen, that the values of $\overline{\alpha}$ and $\overline{\beta}$ for the dyads **1**, **2** and **3** are significantly larger than the individual chromophores **5**, **6** and **7**. Similar behaviour is also observed for $\overline{\gamma}$.

3.2.3 One- and Two-Photon Absorption Spectra

In order to gain an insight into the electronic structure of the investigated systems, the calculations using the configuration interaction with singles method (CIS) based on the INDO approximation were performed with 2000 singly excited configurations included [87] for **1**, **2**, **3** and **4**. This technique was proved to provide reliable prediction of electronic spectra of organic compounds [88]. The theoretically determined one-photon spectra both for **4**, **1**, **2** and **3** is presented in Figs. 3.3, 3.4,

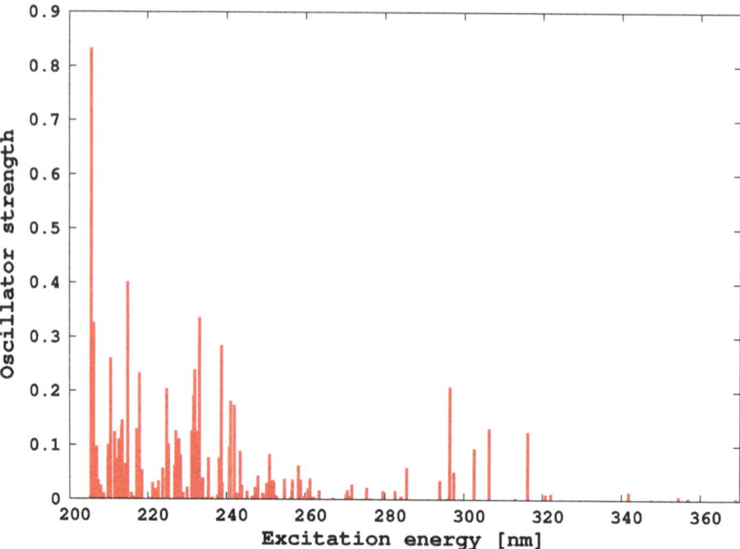

Fig. 3.3 One-photon excitation spectra for **4**

3.5 and 3.6, respectively. The one- and two-photon spectra were calculated without the environmental effects taken into account. The electronic spectrum of **4** measured in cyclohexane was presented by Maggini et al. together with the spectrum of [60]fullerene [89]. In the case of **4**, the authors observed new absorption band present at 306 nm, that was absent in the spectrum of [60]fullerene [89]. This experimental evidence is confirmed by semiempirical calculations reported here, i.e. it is clearly seen from Fig. 3.3, that besides an intense peak at 315 nm, also few intense peaks in the 295–305 nm region appear. The existence of very intense bands below 250 nm is also predicted by results of calculations. In general, the overall agreement between experimental data and results of calculations is satisfactory. The direct comparison of Figs. 3.3 and 3.4 reveals the existence of very intense peak located at 262 nm, missing in the spectrum of **4**. The value of oscillator strength associated with this transition exceeds 0.6. The analysis of the CI vectors reveals that this $\pi \rightarrow \pi^*$ transition takes place in the pyrrolidine ring of carbazole moiety. In the case of compounds **2** and **3** one observes very intense transitions in their spectra at 412 and 418 nm, respectively. The location of absorption band maximum for **3** is in very good agreement with the experimental value 436 nm [39]. The authors attributed this transition to intramolecular charge-transfer from the donor (amine) to acceptor (benzothiadiazole) moiety [39]. It is worth noticing that the range of spectra below 260 nm looks very similar in all four cases and is connected with excitations localized mainly in [60]fullerene moiety. Figure 3.6 contains also the results of calculations of excitation spectra for **3** performed at the B3LYP/3-21G level of theory. It is worth noticing, that the excitation energy to CT state is significantly underestimatedat this level of theory and is

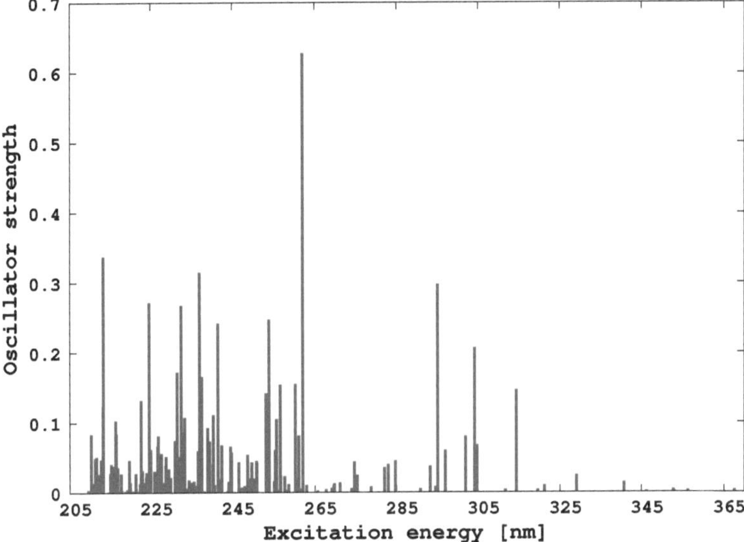

Fig. 3.4 One-photon excitation spectra for **1**

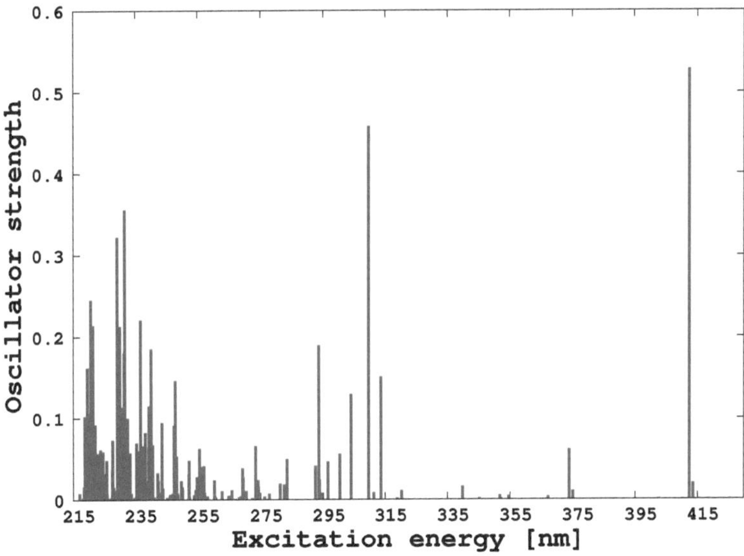

Fig. 3.5 One-photon excitation spectra for **2**

predicted to be 553 nm. The excitation is dominated by one-electron transition from molecular orbital 314 to molecular orbital 318. These molecular orbitals are plotted on Fig. 3.7. As it is seen, this excitation is associated with the density reorganization within benzothiadiazole moiety.

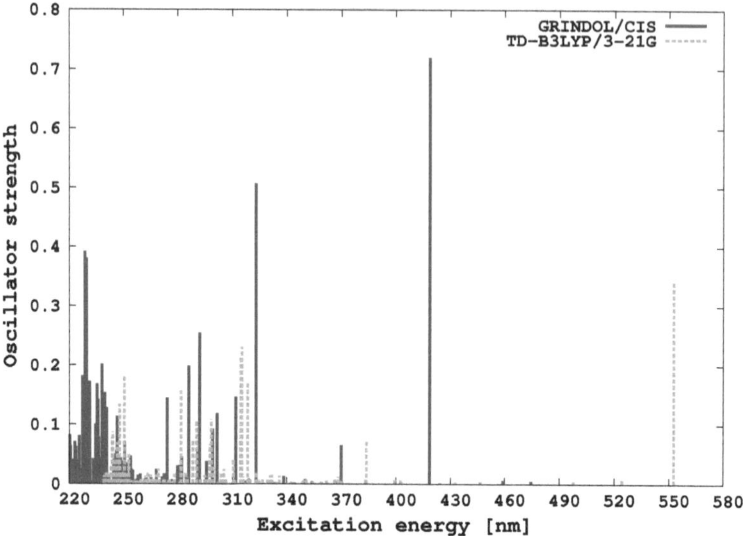

Fig. 3.6 One-photon excitation spectra for **3**

In Table 3.8 we also present the dipole moment differences between the ground and excited states for molecules **1**, **2**, **3** and **4** calculated as expectation values based on CIS wavefunction. We included only the states for which $|\Delta\mu| = |\mu_e - \mu_g| > 2$ D (0.787 au) and f > 0.05, where μ_e and μ_g denote excited and ground state dipole moment, respectively. It is very convenient practice, and commonly followed, to analyze the nonlinear optical properties of molecules in terms of electronic structure of molecules [7, 11]. The values of $|\Delta\mu|$ together with the values of oscillator strengths and excitation energies may provide a qualitative information about the origin of first- [11] and second-order hyperpolarizability [90] as well as two-photon absorption cross section [91, 92]. Thus, heuristic-free structure-property relationships can be established [7]. It follows from Table 3.8 that in some cases (**2** and **3**) there are low-energy, very intense transitions accompanied by $|\Delta\mu|$ value near 3 D (1.18 au). Figure 3.8 presents the dipole moment components of **1**, **2** and **3** in their ground state as well as in selected excited states. The values of total dipole moments for these states are listed in Table 3.8. It is well accepted that [60]fullerene is a moderate electron acceptor [12, 93–95]. It appears, that in all three cases (**1**, **2** and **3**) [60]fullerene serves as an electron withdrawing moiety in the ground electronic state indeed. **1** ([60]fullerene-carbazole) is an example of A–D dyad [96, 97]. It is very interesting to note that in the case of **2** and **3** [60]fullerene exhibits stronger electron withdrawing capability than 2,1,3-benzothiadiazole-based unit [39, 98, 99]. Although [60]fullerene-donor systems have been extensively studied, [60]fullerene-acceptor dyads have been scarcely analyzed [95]. It is known, that [60]fullerene moiety may act as electron-donating group in the excited states [12]. It is seen from Fig. 3.8, that in some excited states of **1**, **2** and **3**, the electron density reorganization upon

Fig. 3.7 Molecular orbitals 314 (*upper plot*) and 318 (*lower plot*) for **3** determined at the B3LYP/3-21G level of theory

Table 3.8 Excitation energies (Δ E) and oscillator strengths (f) of electronic transitions to singlet states for **1**, **2**, **3** and **4**

Label	Δ E [nm]	$\Delta\mu$ [D] (au)	Oscillator strength
		1	
1.	231.6	6.44 (2.53)	0.05
2.	231.5	3.66 (1.44)	0.17
3.	231.4	6.32 (2.49)	0.12
4.	227.5	2.93 (1.15)	0.06
5.	226.8	4.82 (1.90)	0.08
6.	224.9	7.73 (3.04)	0.06
7.	222.5	4.39 (1.73)	0.13
		2	
1.	412.3	2.67 (1.05)	0.52
2.	248.2	2.78 (1.09)	0.05
3.	222.7	5.05 (1.99)	0.06
4.	221.9	4.44 (1.75)	0.07
5.	220.6	2.77 (1.09)	0.24
		3	
1.	418.6	2.92 (1.15)	0.74
2.	246.7	2.33 (0.92)	0.11
3.	246.4	23.51 (9.25)	0.05
4.	231.2	3.42 (1.35)	0.17
5.	229.7	6.67 (2.62)	0.05
6.	229.2	2.23 (0.88)	0.19
7.	227.8	7.69 (3.03)	0.15
8.	227.3	2.33 (0.92)	0.18
9.	223.5	2.77 (1.09)	0.06
		4	
1.	228.4	5.71 (2.25)	0.08
2.	213.2	2.92 (1.15)	0.13
3.	211.5	5.16 (2.03)	0.12

Shown are only the states for which dipole moment difference between a given excited state and the ground state ($\Delta\mu$) is larger than 2.0 D and f > 0.05

excitation changes the direction of dipole moment vector, i.e. the [60]fullerene moiety acts as a electron donor to carbazole-based and 2,1,3-benzothiadiazole-based units. Since **3** exhibit the largest value of β among studied systems, it is interesting to see how the excited states contribute to its value. Figure 3.9 presents the dependence of β_{xxx} as a function of the number of excited electronic states calculated using sum-over-states method with the aid of GRINDOL program. The β_{xxx} for 500 excited states was set to the MP2/3-21G value and all remaining values were scaled accordingly. The number of excited states included in summation refers to the ordering of excited states in Table 3.8, i.e. labeling 1–9 corresponds to the number of states equal to 8,108,110,155,161,163,169,170 and 183, respectively. It follows from Fig. 3.9, that excitation associated with the intramolecular charge transfer within the benzothiazole moiety gives significant contribution to β_{xxx}. It is interesting to note, that the contributions of excited states within the range 8–300 almost perfectly cancels out each other. It can be explained on the basis of data presented in Fig. 3.8, i.e. the orientation of x-component of dipole moment is different for different excited states. However, interpretation of Fig. 3.8 gives only qualitative picture due to the neglect of the coupling between excited states.

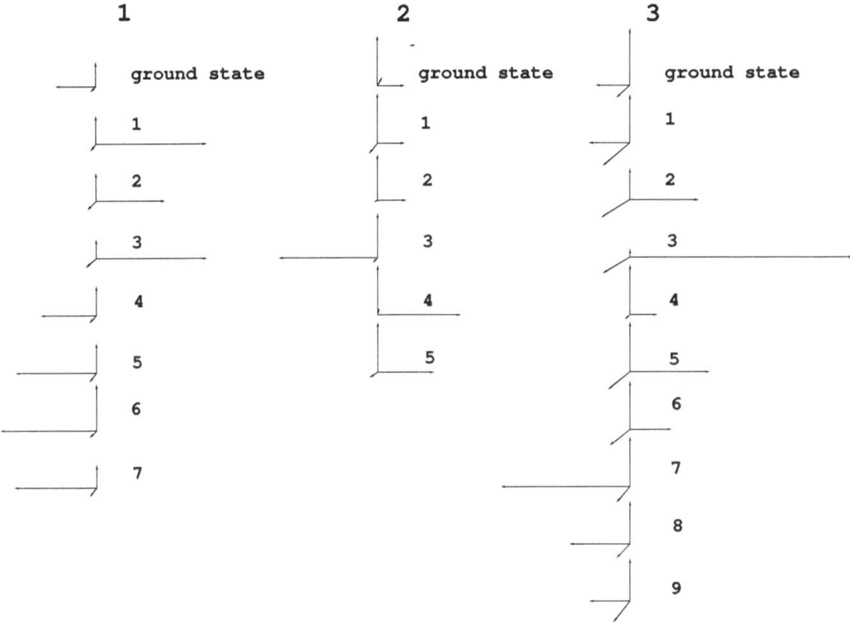

Fig. 3.8 Schematic representation of dipole moment components in the ground and selected excited states for **1**, **2** and **3**, respectively. The values of total dipole moments are given in Table 3.8

Although two-photon absorption phenomenon was predicted by Maria Goppert–Mayer in early 30's of past century [100],it was quite recently that it has attracted significant attention of the scientific community [101–109]. This is certainly due to the various optical applications, which include optical limiting [110] and optical data storage [111]. Based on numerous theoretical and experimental activities, it was possible to establish structure-property relationships for organic conjugated molecules [101, 102, 108]. However, only a small number of studies were devoted to theoretical analysis of two-photon absorption spectra of functionalized fullerenes [112]. It was observed experimentally, that values of two-photon absorption cross section for [60]fullerene-chromophore dyads can be as large as 10^4 GM (1 GM $= 1 \cdot 10^{-50}$ $\mathrm{cm}^4 \cdots \cdot \mathrm{photon}^{-1} \cdot \mathrm{molecule}^{-1}$) in the nanosecond region [113]. However, the measurements were performed only at 800 nm and the observed two-photon absorption can be attributed to excitations within the chromophore moiety [113]. Hence, it seems to be very interesting to analyze TPA activity of **1**, **2**, **3** and **4** on purely theoretical basis in a wide spectral range.

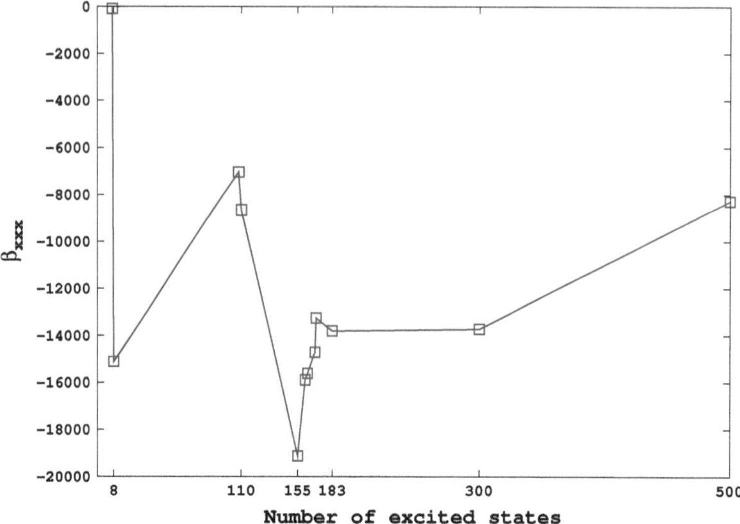

Fig. 3.9 Convergence of β_{xxx} for **3** as a function of the number of excited states includes in sum over states expression

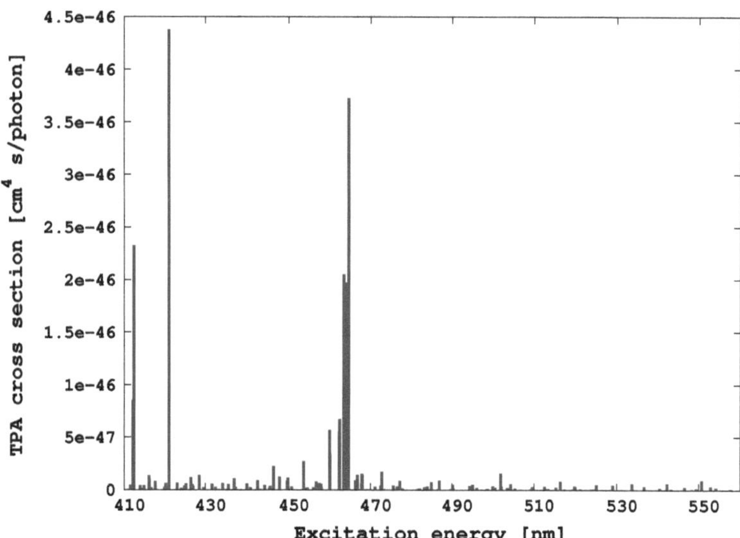

Fig. 3.10 Two-photon absorption spectra for **4**. TPA cross section calculated for linearly polarized photons with parallel orientation

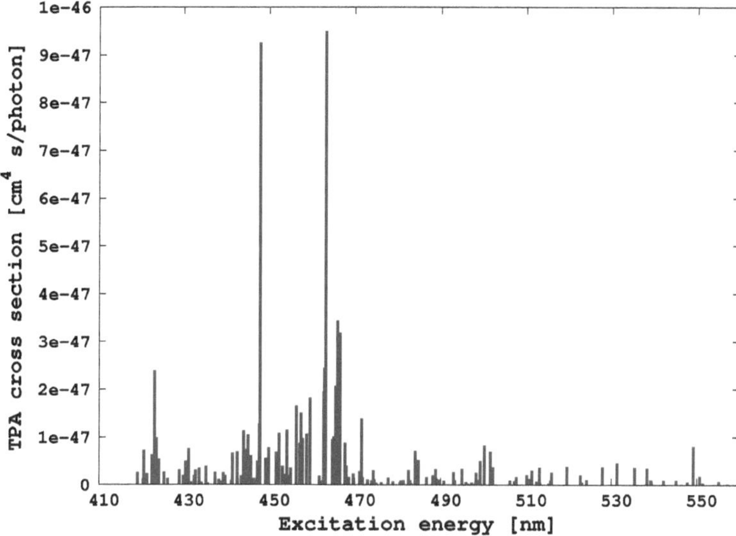

Fig. 3.11 Two-photon absorption spectra for **1**. TPA cross section calculated for linearly polarized photons with parallel orientation

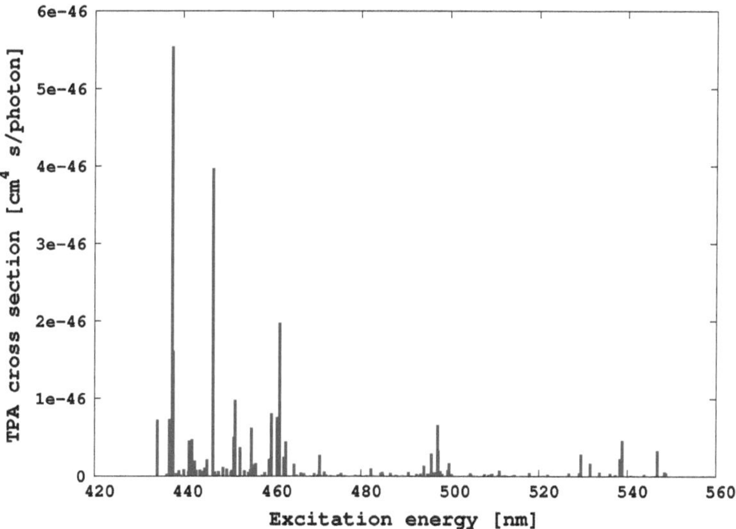

Fig. 3.12 Two-photon absorption spectra for **2**. TPA cross section calculated for linearly polarized photons with parallel orientation

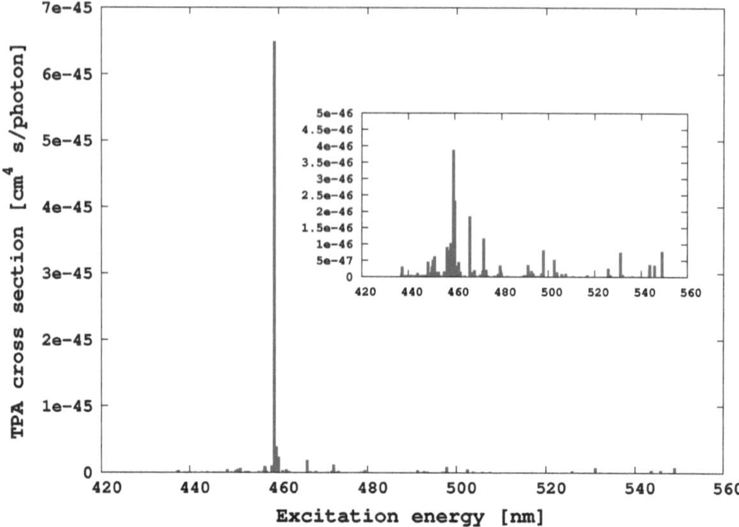

Fig. 3.13 Two-photon absorption spectra for **3**. TPA cross section calculated for linearly polarized photons with parallel orientation. The inset figure presents the TPA spectrum with the exception of the most intense transition

The two-photon absorption spectra of the investigated molecules is presented in Fig. 3.10, 3.11, 3.12 and 3.13. The plots were prepared assuming the value of line shape function equal $5{\cdot}10^{-15}$ 1/s. This choice was verified by test calculations for 2,1,3-benzothiadiazole derivative for which experimental value of two-photon absorption cross section is available [98] . As it is seen, the largest values of two-photon absorption cross sections appears in the region 400–500 nm, where one-photon absorption of [60]fullerene moiety appear. The values of two-photon absorption cross sections are of the order of 10^4 GM. Based on the results from Figs. 3.10, 3.11, 3.12 and 3.13, we can conclude that two-photon absorption cross section for **3** is almost an order of magnitude larger than that for remaining studied systems in the 400–500 nm region. It is consistent with the values of electronic contributions to Reγ: The $\overline{\gamma}$ value for **3** is few times larger than the corresponding value of **1** and **2** (see Table 3.6).

3.2.4 Vibrational Contributions to Electric Polarizabilities

In this section, we shall present the results of calculations of vibrational contributions to α and β performed at the HF/6-31G level of theory for **1**, **2**, **3** (Fig. 3.1) and **11** (Fig. 3.5). Due to the computational limitations, the nuclear relaxation (NR) contributions were evaluated only for **11**. Since the diagonal components (along cartesian x axis) of electronic contributions to α and β were found to be dominant, we

Table 3.9 Vibrational contributions to molecular polarizabilities calculated within the double-harmonic oscillator approximation at the HF/6–31G level of theory for **1**, **2**, **3** and **11**

	$\overline{[\mu^2]}^{(0,0)}$	$\overline{[\mu\alpha]}^{(0,0)}$
1	29.2	5150.0
2	42.2	9256.5
3	119.0	7335.6
11	11.8	268.1

All values are given in atomic units

Table 3.10 Experimentally determined values of the second-order hyperpolarizability of the fullerene derivatives **1–3**

Sample	$Re(\gamma)[\times 10^{-33} esu]$	$Im(\gamma)[\times 10^{-33} esu]$	$\gamma[\times 10^{-33} esu]$
1	-10 ± 2	24 ± 3	26 ± 4
2	-21 ± 6	33 ± 7	38 ± 9
3	69 ± 11	159 ± 21	172 ± 23
* C_{60}	–	28 ± 1	28 ± 1

*Ref. [157]

report on NR vibrational contributions for only diagonal (x) components of the tensors. Moreover, it should be pointed out that NR contributions were calculated fully numerically. In the case of **1**, **2** and **3**, the vibrational contributions were calculated using double-harmonic oscillator approximation [49].

It was found for **11** that $\alpha^{nr}_{xx} = 11.78$ a.u. and is significantly smaller than the electronic counterpart (501.28 a.u.). However, the opposite trend was observed in the case of the static first-order hyperpolarizability. The electronic β^{e}_{xxx} was found to be -35.8 a.u., while $\beta^{nr}_{xxx} = -608.7$ a.u. Thus, the ratio $\beta^{nr}_{xxx}/\beta^{e}_{xxx}$ is equal to 17. In order to explain the origin of the large β^{nr}_{xxx} value, the analysis of contributing terms was performed. The value of the harmonic term $[\mu\alpha]^{(0,0)}$ is 128.7 a.u. which is 3.6 times larger than the electronic contribution to first-order hyperpolarizability. Undoubtedly, the most important contribution comes from the anharmonic terms $[\mu^3]^{(0,1)}$ and $[\mu^3]^{(1,0)}$, i.e. the sum of these two terms is equal to -737.4 a.u. Thus, one may infer that the anharmonicity plays a crucial role in vibrational nonlinear optical response in the case of **11** (accounts for 82.5 % of the total β^{nr}_{xxx} value). Adopting the infinite frequency approximation, which was proven to give reliable results [64], the calculated $\beta^{nr}(-\omega;\omega,0)_{\omega\to\infty}$ value was found to be 42.9 a.u., while the $\gamma^{nr}(-2\omega;\omega,\omega,0)_{\omega\to\infty}$=$-40$ a.u.

In Table 3.9, the vibrational contributions for **11**, **1**, **2** and **3**, calculated within the double-harmonic oscillator approximation are presented. The calculations were performed seminumerically, i.e. second derivatives of energy were calculated by differentiation of analytic first derivatives. Thus, the values of harmonic terms presented in Table 3.9 may serve as a reference point for numerical accuracy assessment of NR contributions discussed above. The relative error for $[\mu\alpha]^{(0,0)}$ term for **11** does not exceed 10 %. It follows from Table 3.9 that diagonal vibrational contributions to α

for **1** and **2** are of similar magnitude. However, the average vibrational polarizability is about 30 % larger for **2**. Both diagonal and average vibrational polarizability of **3** is few times larger than that of **1** and **2**. Likewise, largest vibrational contributions to β are observed for **3**. The $[\alpha^2]^{0,0}_{xxxx}$ term contributing to γ is also largest for **3**. The values of double-harmonic vibrational contributions are of similar magnitude as the electronic counterpart (see Table 3.6). The effect of electron correlation has not been considered in the present study. However, electron correlation may have significant influence on the vibrational linear and nonlinear optical properties.

3.3 Conclusions

In this work we have performed an extensive analysis of electronic and vibrational contributions to molecular polarizabilities for a series of [60]fullerene derivatives. It has been found that modification of [60]fullerene-benzothiazole system by adding triphenylamine moiety leads to three and two times larger value of first- and second-order hyperpolarizability of the whole dyad, respectively. It has also been observed, that the imaginary part of second-order hyperpolarizability is significantly enhanced in the 400–500 nm spectral region. The analysis of nature of excitations for considered systems confirms that [60]fullerene moiety acts as electron donor in excited states, while in the ground state it is a strong electron acceptor. The vibrational contributions to β for [60]fullerene-chromophore dyads, calculated within the double harmonic oscillator approximation, have been found to be much larger than the electronic counterpart. The computations of nuclear relaxation contributions to α and β for N-methylpyrrolidine derivative of [60]fullerene have shown, however, that anharmonicity plays a crucial role in determination of vibrational counterpart of the nonlinear optical response. It should be highlighted that the calculations of anharmonic vibrational contributions to β are reported for the first time for organofullerenes.

A hierarchy of methods including *ab initio* (HF, MP2) and more approximate semiempirical techniques (PM3 and PM6) have been employed. Linear scaling approaches have also been used to lower the computational cost. The assessment of performance of various functionals in calculations of first-order hyperpolarizability has also been performed. The results of calculations reveal that except the long-range density functional theory, all employed functionals significantly overestimate the value of β. In few cases, we have observed that the discrepancy may exceed order of magnitude. The analysis of first-order hyperpolarizability of several model systems, composed of [60]fullerene and substitutes of different electron-donating capabilities, reveals that the so-called 'DFT catastrophe' is certainly connected with the donor–acceptor character of investigated systems. On the other hand, both LC-BLYP and CAM-B3LYP functionals provide reliable prediction of β values for the investigated [60]fullerene-chromophore dyads. Thus, density functional theory accounting for long-range effects may serve as an reliable tool for the calculation of electronic contributions to nonlinear optical properties for organofullerenes.

3.4 Triphenylamine-Functionalized Fullerenes

Aiming at design of effective solar energy harvesting devices, many fullerene derivatives have been studied during last decade [124–127]. The interest in this class of compounds is due to the high quantum efficiency of photoinduced electron transfer which occurs in the femtosecond time scale. The back-donation of electron is several orders of magnitude slower, which makes C_{60} dyads promising candidates for photovoltaic devices. On the other hand, the interest in C_{60}-chromophore systems is also due to their potential applications in nonlinear optics [128–131]. For these reasons, photophysical properties of fullerene derivatives are being extensively studied [132–136]. As it was repeatedly pointed out by several authors [129, 132], high efficiency of intersystem crossing between the singlet and the triplet excited state can be achieved in the case of C_{60}. It might be further increased by the heavy-atom effect [131]. Moreover, the microwave irradiation used in microwave induced absorption (MIA) can cause the redistribution of population of the triplet sublevels which might amplify singlet-triplet conversion [136]. Significant population in the triplet excited state after photoexcitation increases the probability of excited state absorption (ESA). As a result of large population in the triplet excited state, the reverse saturable absorption (RSA) can be observed. This phenomenon is considered nowadays to be superior to the two-photon absorption (TPA) process as far as optical-limiting applications are concerned. TPA, however, might be used to gain the RSA state of question. A combined TPA-RSA approach was indeed successfully applied to diphenylaminofluorene-C_{60} dyads and triads [113]. In order to fully exploit nonlinear optical properties within the telecommunication range, a detailed characteristics of photophysical properties of materials is necessary, just to mention one- and two-photon absorption spectra, intersystem crossing rates or fluorescence quantum yields.

TPhA-based materials are widely known as excellent hole-transporters and electroluminescent components [137–139], while films of TPhA have been used in organic-light-emitting-diodes (OLEDs), and in opto-electronic materials owing to the high electron-donor ability and good film-forming properties [140]. Due to deviation of the three phenyl rings from co-planarity, TPhA materials can be considered as 3D systems. Additionally, TPhA is characterized by an amorphous state potentially useful, when combined with novel conjugated electron-acceptor systems, for the development of advanced material architectures with isotropic optical and charge-transport properties. On the basis of all the above, we focused on the preparation of a class of push-pull hybrid materials based on TPhA and C_{60} (see Fig. 3.14). It is expected that the electronic push-pull interactions would not only help charge transfer from TPhA to C_{60}, but also afford the photorefractivity by electrooptic effect. Triphenylamine plays the dual role of an electron donor and a photoconductor for photorefractivity while the fullerene sphere is known to exhibit a sizable nonlinear optical (NLO) response [141] due to their large π-conjugated surface and the extensive charge delocalization [142]. Moreover, in order to reveal the effect of a second fullerene to the C_{60}-donor dyads, related triads consisting of two

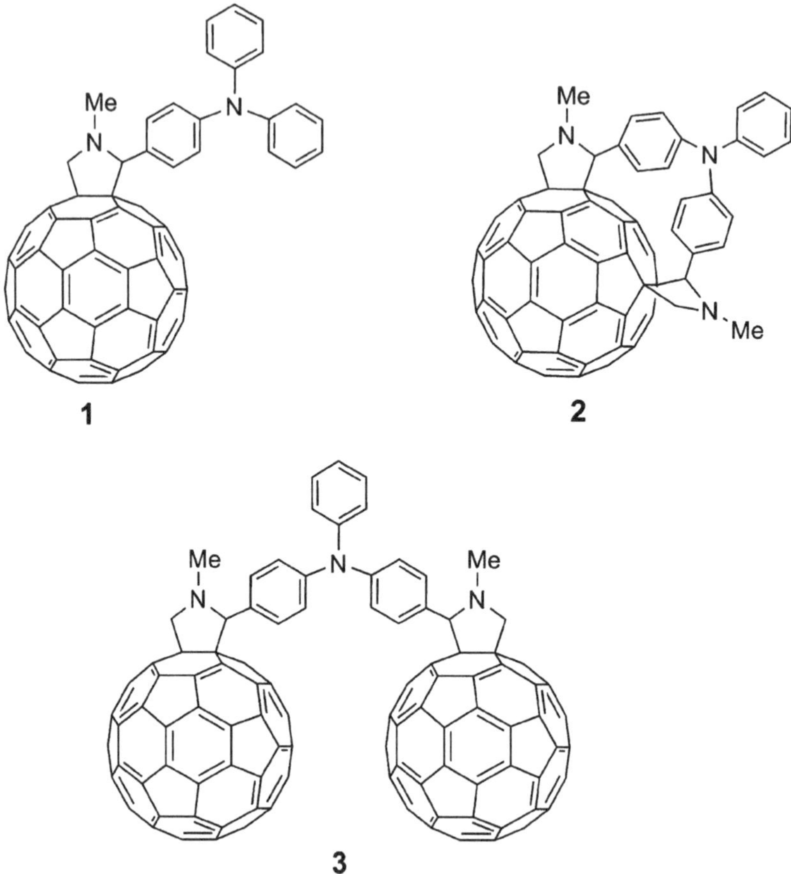

Fig. 3.14 TPhA-C_{60} **1**, bisadduct TPhA-C_{60} **2** and dumbell TPhA-$(C_{60})_2$ **3** hybrid materials investigated in the present study

C_{60} units covalently linked with a TPhA unit were prepared and investigated. The presence of two C_{60} units could contribute to establish well-ordered solid phases, which is a crucial factor in optoelectronic devices [143].

The aim of the current work is three-fold, namely (i) the synthesis of the mono- and bis- triphenylamine adducts of C_{60} as well as the dumbell TPhA-bridged-$(C_{60})_2$ hybrid, (ii) the evaluation of the third-order nonlinear optical properties of the TPhA-modified C_{60}-based hybrid materials, and (iii) the theoretical calculations aiming at better understanding of the structure-property relations for the investigated push-pull hybrid materials.

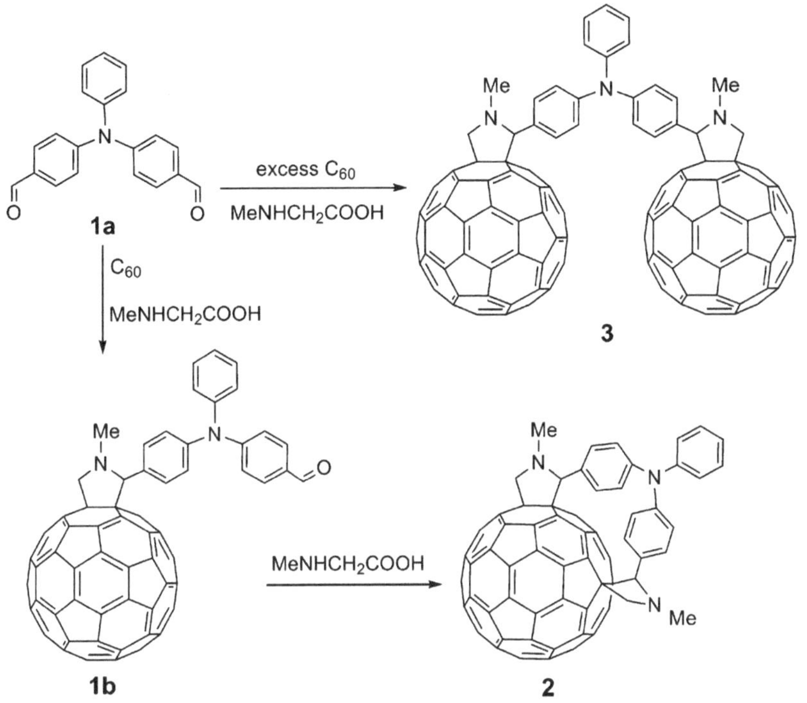

Fig. 3.15 Synthetic route furnishing bis-adduct TPhA-C_{60} **2** and dumbell TPhA-$(C_{60})_2$ **3** hybrid materials

3.4.1 Synthesis

Mono-adduct **1** was prepared in 43 % yield using a modified published procedure, [137] where C_{60}, 4-(diphenylamino)benzaldehyde and sarcosine were stirred in refluxing toluene overnight. The regioselective synthesis of the equatorial C_{60} bis-adduct **2** was accomplished in a two-step mono-addition reaction sequence of azomethine ylides onto C_{60}, in which triphenylamine itself is utilized as a rigid tether [144]. In the first step, 1,3-dipolar cycloaddition of azomethine ylides, generated in-situ upon thermal condensation of TPhA bis-aldehyde [145] **1a** and sarcosine onto C_{60}, resulted in the mono-fulleropyrrolidine adduct **1b** having a free aldehyde unit. In the following step, further reaction with sarcosine resulted in the formation of the second fused pyrrolidine ring onto the fullerene sphere, with complete regioselectivity, at the equatorial position, to furnish bis-adduct **2** (see Fig. 3.15).

The synthesis of a bis-fullerene material **3**, where two pyrrolidine functionalized carbon spheres are bridged together via a rigid TPhA unit was achieved in one step [146]. The reaction conditions followed were important in order to obtain bis-fullerene material **3** in good yield. Such optimized reaction parametres include the utility of (i) excess C_{60}, (ii) small reaction volume and (iii) short reaction time.

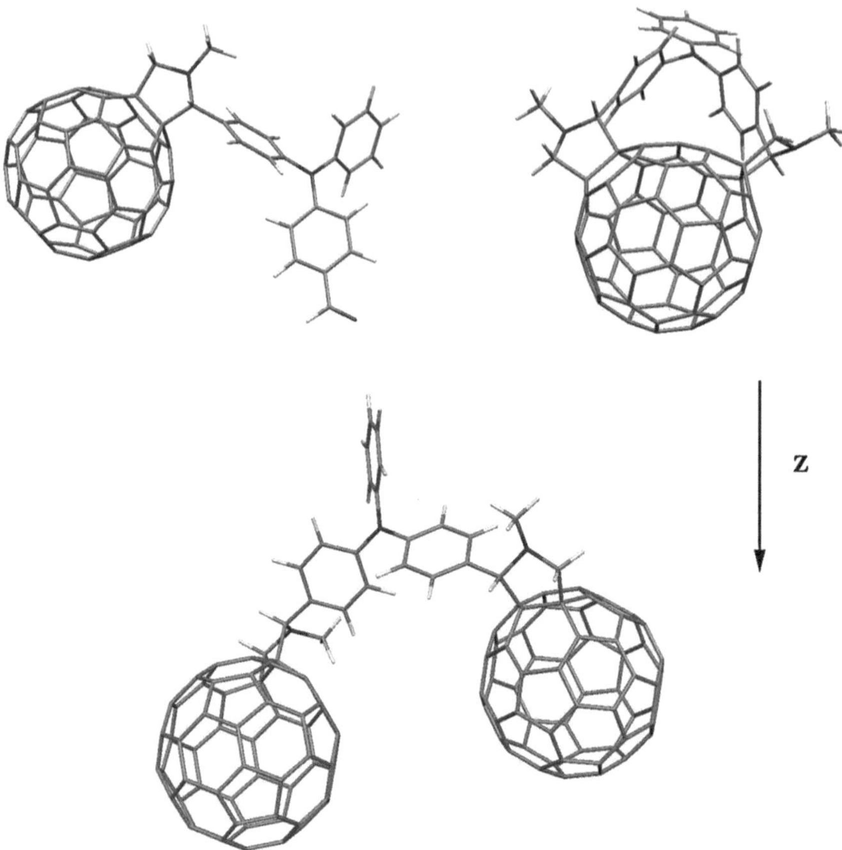

Fig. 3.16 Orientation of **1b** (*top left*), **2** (*top right*) and **3** (*bottom*) in Cartesian coordinates used for property computation

In this context, dialdehyde **1a** (5 mg, 0.017 mmol), sarcosine (14 mg, 0.157 mmol) and five-fold excess of C_{60} (60 mg, 0.083 mmol) dissolved in 10 ml of o-dichlorobenzene (o-DCB) heated at 130 °C for 90 min. After a typical work up procedure, the reaction mixture is purified by recycling HPLC on a preparative Buckyprep Cosmosil column (250 × 20 mm, toluene eluent, 10 ml/min flow rate) to furnish bis-fullerene derivative **3** in 32 % yield. The bis-fullerene **3** shows moderate solubility, thus allowed us to record ^{1}H and ^{13}C NMR spectra (Supp. Info., Fig. S1) corroborating the depicted structure in Scheme 1. The attenuated-total-reflectance infra-red (ATR-IR) spectrum of 3 demonstrates the characteristic stretching vibrations of the C-H (2803–3030 cm^{-1}) as well as the characteristic absorptions for fullerenes (Supp. Info., Fig. S2). Matrix-assisted-laser-desorption time-of-flight mass-spectroscopy (MALDI-TOF-MS), in the negative ionization mode and with the aid of trans-2-[3-(4-tert-butylphenyl)-2-methyl-2-propenylidene]malononitrile shows the molecular ion peak of 3 accompanied with a strong fragmentation at m/z 720 corresponding to C_{60} (Supp. Info., Fig. S3).

3.4.2 Theoretical Methods

In the present study, the quantity of primary interest is the second-order hyperpolarizability (γ) which is defined as one of the coefficients in the Taylor expansion of the total energy of a molecule in the presence of the electric field [65]. In this study we use Eq. 3.1 to compute real part of γ by numerical differentiation of the total energy with respect to the electric field [147]. Thus, at no additional cost β can also be determined. In order to make a valid comparison for the three investigated systems, we use the orientationally averaged molecular (hyper)polarizabilities [65]. The geometries of the investigated systems were optimized at the B3LYP/6-31G(d) level of theory. The hessian was also computed in order to confirm that the structures correspond to minima on the potential energy hypersurface. Due to the computational restrictions, at certain levels of theory only diagonal components of the β and γ were determined. For this reason all the structures were oriented in the Cartesian coordinates in such a way to make the dipole moment vector parallel to the Cartesian z-axis. The orientation of the studied molecules used for property evaluation is presented in Fig 3.16. All the results of calculations presented in this study were obtained with the aid of the Gaussian 03 [148] and GAMESS US [149] programs, except the computations using CAM-B3LYP functional. In that event we used the modified version of the DALTON program [150].

3.4.3 Nonlinear Optical Measurements

For the measurements of the NLO properties of organofullerenes 1,2 and 3, the Z-scan technique has been employed [151]. Z-scan is a relatively simple experimental technique allowing for the simultaneous determination of the real and imaginary parts of the third-order susceptibility $\chi^{(3)}$. Then, the second-order hyperpolarisability γ can be obtained, which is a molecular constant, as it does not depend on the concentration and characterizes the nonlinear response of the molecule. Briefly, in the Z-scan technique, the transmittance of a sample is measured under two different experimental configurations as it moves along the propagation direction of a focused Gaussian laser beam, therefore experiencing different intensity at each position. So, the transmitted through the sample laser beam is totally collected after the sample by means of a large diameter focusing lens (i.e. open-aperture Z-scan); the transmitted laser beam is also simultaneously measured after the sample and after passing through a narrow aperture placed in the far field (i.e. closed-aperture Z-scan). From the open-aperture Z-scan the nonlinear absorption β of the sample is determined, while from the division of the closed-aperture Z-scan trace by the open-aperture Z-scan trace,

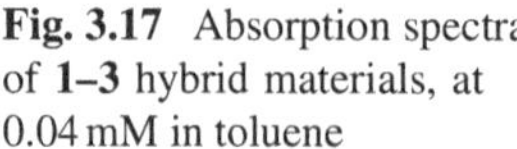
Fig. 3.17 Absorption spectra of **1–3** hybrid materials, at 0.04 mM in toluene

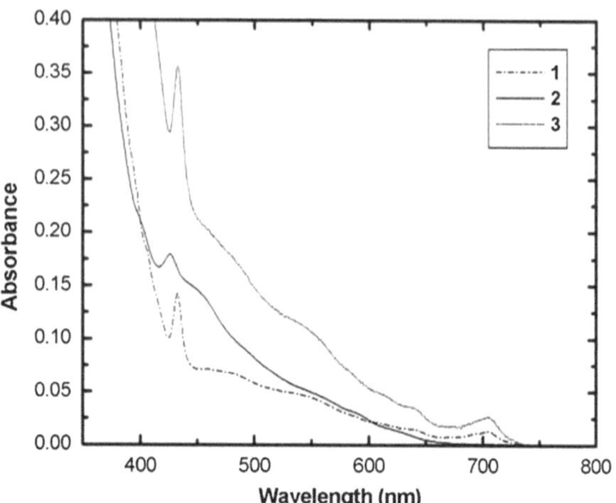

the so called divided Z-scan is obtained, which allows for the determination of the nonlinear refraction parameter γ' of the sample. The former quantity is related to the imaginary part of the third-order susceptibility ($\mathrm{Im}\chi^{(3)}$), while the latter to the real part of the third-order susceptibility. The open-aperture Z-scan trace may exhibit a transmittance valley or peak near the focal plane indicative of positive or negative imaginary part of the third-order nonlinearity respectively. In the former case the sample exhibits reverse saturable absorption (RSA) behavior, while in the latter case, it exhibits saturable absorption (SA) behavior. On the other hand the divided Z-scan might exhibit either a valley-peak or peak-valley configuration, indicative of positive or negative nonlinear refractive index (n_2, $\mathrm{Re}\chi^{(3)}$) respectively, the sample acting as a positive (focusing) lens or negative (defocusing) lens.

In the present study, the second harmonic at 532 nm of a 35 ps mode-locked Nd:YAG laser operating at a repetition rate of 10 Hz has been employed. The spatial and the temporal profile of the laser beam were characterized and found to have a Gaussian distribution. For the needs of the measurements several solutions in toluene have been prepared having different concentrations for each fullerene system. The UV-Vis-NIR absorption spectra of the prepared solutions were recorded before and after the measurements to ensure that no photodegradation due to laser irradiation has occurred.

3.4.4 One-Photon Absorption Spectra

In Fig. 3.17 the UV-Vis-NIR spectra of the studied fullerene derivatives are presented. The spectra of bis-fullerene material **3** shows the characteristic absorptions of a fulleropyrrolidine monoadduct at 433 nm as well as that of TPhA at the UV region (see Figs. 3.17 and 3.18). The assignment of the band located at 433 nm has

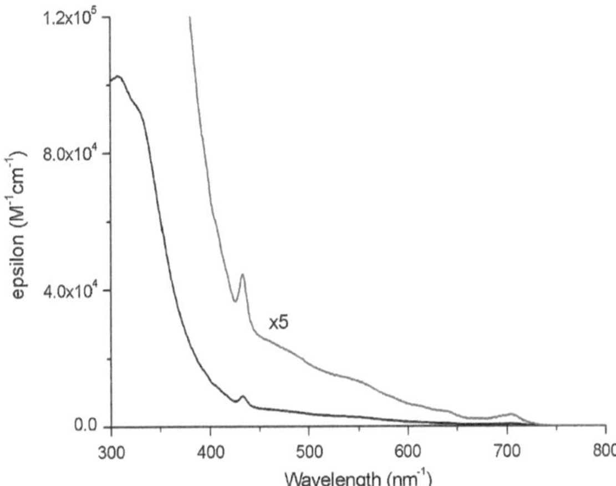

Fig. 3.18 Absorption spectrum of dumbell TPhA-$(C_{60})_2$ **3** in toluene

been the subject of long discussion [152]. Importantly, the absorption spectrum of **3** is considered as the superimposition of the absorption spectra of a fulleropyrrolidine monoadduct and TPhA, regardless of the solvent used to obtain the spectra (i.e. toluene, chloroform and DMF). Therefore, it is reasonable to assume that there are not appreciable and strong interactions in the ground state between the spheroid units and the bridge component in the TPhA-linked bis-fullerenes. For this reason, we have not attempted to determine the electronic spectra of **3** computationally. In order to simulate the spectra of **1b** and **2** presented in Fig. 3.19, we used the PBE0 functional which was proven in the past to reliably predict the excitation energies of organic π-conjugated organic chromophores [153, 154]. The spectrum of **1b** in the region 200–280 nm is similar to that of N-methylfulleropyrrolidine measured in cyclohexane by Maggini et al. [155] In the case of **1b**, however, two quite intense bands appear at 293 nm and 340 nm. The results of time-dependent density functional theory (TDDFT) calculations show that the leading single-excited configurations are HOMO→LUMO+7 and HOMO→LUMO+15 for excitations at 340 and 293 nm, respectively. In both cases, the excitation is well localized within the triphenylamine moiety and does not involve the orbitals localized on [60]fullerene. The excitations below 300 nm can be ascribed to transitions within the fullerropyrolidine moiety [155]. The spectra of **2** is quite different from that simulated for **1b**. Firstly, one does not observe the intense transitions at 340 and 290 nm. The excitations in this region are predicted for **2** to be much less intense than for **1b**. The oscillator strength for the transition at 340 nm does not exceed 0.05. The analysis of the nature of this excitation reveals strong mixing of two single excited configurations, namely HOMO→LUMO+5 and HOMO→LUMO+8. The former involves charge transfer from the triphenylamine to the [60]fullerene, while the latter is localized solely within the triphenylamine moiety (see Fig. 3.19). The experimental spectra of *N*-methylfulleropyrrolidine reveals the existence of two distinct bands located near 210 and 250 nm [155]. These bands are also observed in the theoretically simulated

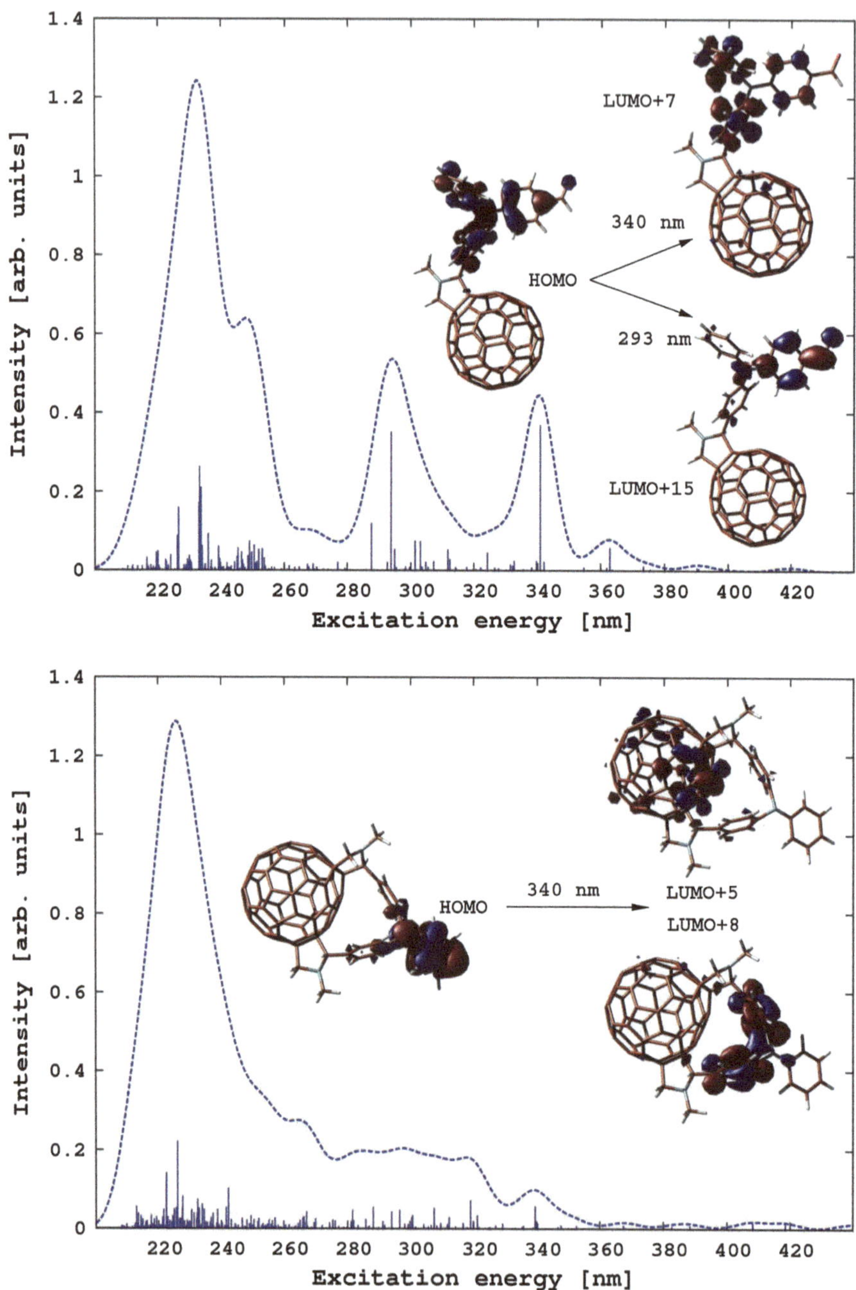

Fig. 3.19 One-photon absorption spectra of **1** (*top*) and **2** (*bottom*) simulated at the PBE0/6-31G(d) level of theory

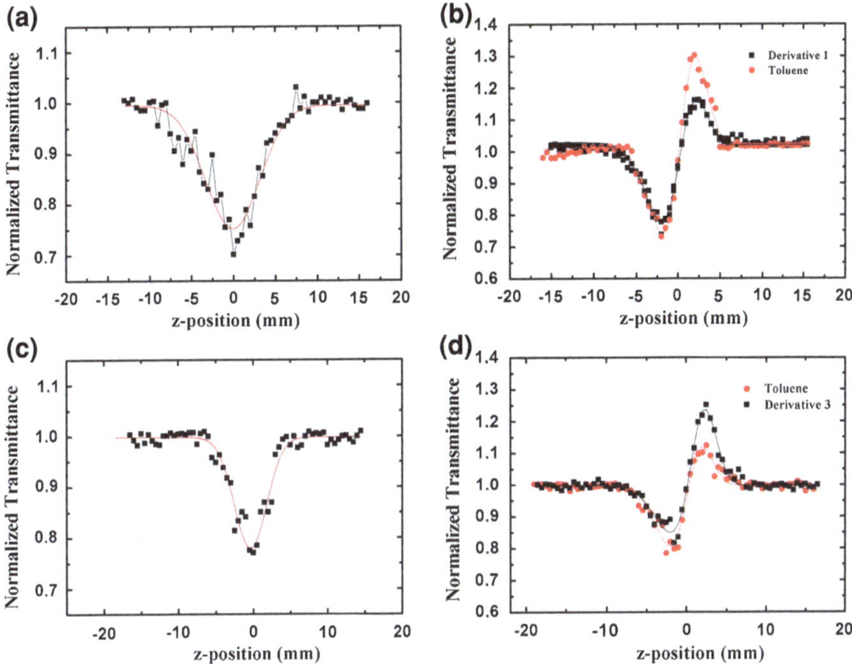

Fig. 3.20 Open **a**, **c** and divided **b**, **d** Z-scans of 1.24 and 0.195 mM toluene solutions of derivatives **1** (A, B) and **3** obtained at incident energies of 3.8 and 2.12 μJ corresponding to peak intensities of 21.3 and 11.9 GW/cm^{-2}

spectra of **1b**. The feature at 250 nm, however, is not clearly visible in the simulated spectra of **2**.

3.4.5 Nonlinear Optical Properties

In (Fig. 3.20), some characteristic Z-scan recordings of derivatives **1** and **3** respectively dissolved in toluene are presented as an example. Graphs 3.20a and 3.20b, show the open-aperture and divided Z-scan traces of a 1.24 mM toluene solution of derivative 1, while graphs 3.20c and 3.20d present similar Z-scans of a 0.195 mM toluene solution of derivative **3**. As can be seen, the open-aperture Z-scans of both derivatives exhibit a characteristic transmittance valley near the focal plane, indicative of reverse saturable absorption (RSA) behavior, while the corresponding divided Z-scan traces present a valley-peak configuration. The nonlinear absorption of toluene was measured separately under identical experimental conditions and it was found negligible. Therefore, the measured open aperture Z-scans were revealing straightforwardly the absorptive nonlinearity of the fullerene derivative molecules.

On the contrary, toluene was found to exhibit significant refractive nonlinearity under the experimental conditions used, as it is clearly shown in the divided Z-scans of neat toluene which exhibit a valley-peak configuration indicative of a focusing behavior (i.e positive $\mathrm{Re}\chi^{(3)}$) in agreement with other works [156]. Some representative divided Z-scan traces of the derivatives **1**, **3** and toluene measured under identical experimental conditions are shown in Fig. 3.20b,d. It is obvious that in the case of derivative **1** the difference between the valley and the peak is lower than that of neat toluene, suggesting defocusing behavior of the derivative **1**. On the other hand the ΔTp-v of derivative **3** is larger than that of toluene, so it exhibits focusing behavior, i.e. the two fullerene derivatives exhibit opposite sign nonlinearity. The fullerene derivative **2** was found to exhibit a behavior similar to that of derivative **1**. Then, the Δ Tp-v values were plotted as a function of the incident laser energy, for the different concentration solutions of molecules **1**, **2** and **3** (Fig. 3.21). The solid lines correspond to the best fit of the experimental data. As can be seen, the slopes of the Δ Tp-v values are decreasing increasing the concentration in the case of fullerene derivative **1**, indicating the opposite sign refractive nonlinearity between fullerene molecule **1** and toluene, while the opposite is observed for the molecule **3**. Considering that toluene exhibits focusing behavior under 35 ps, 532 nm excitation, it becomes evident that the derivatives **1** and **2** exhibit (similar) defocusing behavior (i.e. negative $\mathrm{Re}\chi^{(3)}$) while molecule **3** exhibits focusing behavior (i.e. positive $\mathrm{Re}\chi^{(3)}$).

Then, the nonlinear refractive parameter γ' have been determined according to the following relations [151]:

$$T = 1 - \frac{4\Delta\Phi_0 x}{(x^2+9)(x^2+1)} \tag{3.20}$$

$$\Delta\Phi_0 = kI_0\gamma' L_{eff} \tag{3.21}$$

where $\Delta\Phi_0$ is the on-axis nonlinear phase shift at the focus, I_0 is the peak intensity of the laser pulse at the focal plane. L_{eff} is given by the equation:

$$L_{eff} = [1 - exp(-a_0 L)]/a_0 \tag{3.22}$$

where a_0 is the linear absorption coefficient and L the optical thickness of the investigated sample. Accordingly, the nonlinear absorption coefficient β was determined from the open-aperture Z-scans using the following equation [151]:

$$T(z) = \sum_{m=0}^{\infty} \frac{[-q_0(z,0)]^m}{(m+1)^{3/2}} \tag{3.23}$$

where q_0(z,0) is defined from the following equation:

$$q_0(z,0) = \beta I_0 L_{eff}/(1 + z^2/z_0^2) \tag{3.24}$$

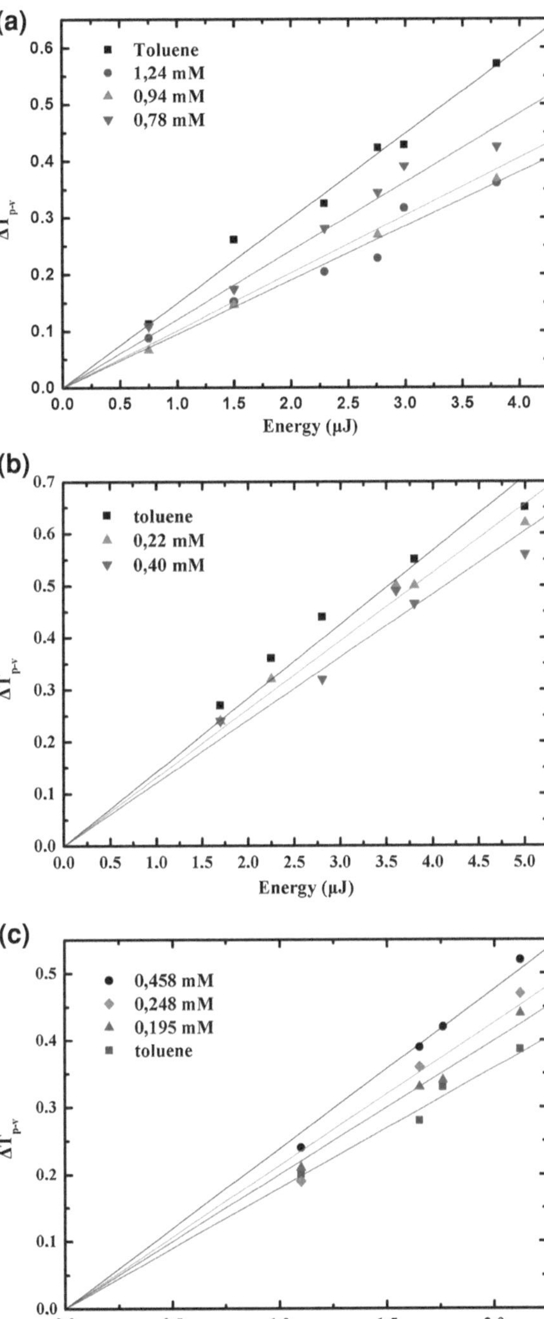

Fig. 3.21 Variation of the Δ Tp-v as a function of the incident laser energy. **a** Derivative **1**, **b** derivative **2**, **c** derivative **3**

where β is the nonlinear absorption coefficient and z_0 is the Rayleigh range of the laser beam.

From the determined γ' and β values, the real and imaginary parts of the third-order susceptibility $\chi^{(3)}$, i.e. Re$\chi^{(3)}$ and Im$\chi^{(3)}$, can be obtained using the following relations [151]:

$$Re[\chi^{(3)}] = 2n_0^2\varepsilon_0 c\gamma \tag{3.25}$$

$$Im[\chi^{(3)}] = \frac{n_0^2\varepsilon_0 c^2}{\omega}\beta \tag{3.26}$$

where n_0 is the refractive index of the solvent, ε_0 is the electric permeability, c is the speed of light in m sec^{-1} and ω is the cyclic frequency of the incident laser beam. The nonlinear refractive parameter γ' (m^2W^{-1}) and the nonlinear refractive index n_2 (esu) are related through the relation [151]:

$$n_2(esu) = (\frac{c(m/s)n_0}{40\pi})\gamma' \tag{3.27}$$

Finally, the second-order hyperpolarizability γ has been deduced according to the equation:

$$\gamma = \frac{\chi^{(3)}}{NL^4} \tag{3.28}$$

where N is the number density and L is the local field correction factor. Detailed description of the analysis for the determination of the nonlinear optical parameters from the experimental data can be found elsewhere [151].

The obtained values of the second-order hyperpolarizability of the fullerene derivatives **1**, **2** and **3** are presented in Table 3.11. In order to facilitate direct comparison between the results concerning the fullerene derivatives **1**, **2**, **3** and C_{60}, some C_{60} toluene solutions have been also studied under identical experimental conditions. C_{60} has been found to exhibit negligible nonlinear refraction and the nonlinear refractive efficiency of the solutions is attributed to the solvent (toluene). On the other hand the nonlinear absorption of the neat fullerene has been found to be significant. In Fig. 3.22a characteristic open aperture Z-scan of a C_{60}-toluene solution can be seen. The obtained nonlinear optical parameters can be also seen in Table 3.11. It is obvious that all fullerene derivatives **1–3** exhibit much higher nonlinear refraction than that of C_{60}. Moreover, while the nonlinear absorption of derivatives **1** and **2** is comparable with that of C_{60}, the derivative **3** was found to exhibit an almost five times larger nonlinear absorption than that of C_{60}. The results are in agreement with what has been reported by Albota et al. [158] They studied several donor-π-donor, donor-acceptor-donor, and acceptor-donor-acceptor compounds and observed that symmetric charge transfer upon excitation leads to enhancement of two-photon absorption cross section. The present C_{60} results are in good agreement with previously reported results measured under similar experimental conditions (i.e. ps laser pulses). For example Mavritsky et al. [159] using 10 ps pulses at 527.5 nm found an Imγ of 120×10^{-33}

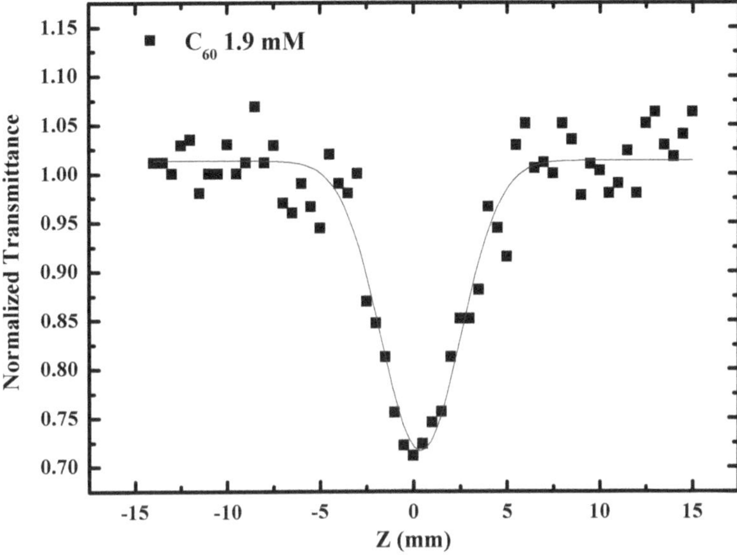

Fig. 3.22 Open-aperture Z-scan of a 1.9 mM C_{60} toluene solution (1,5 μJ, 532 nm, 35 ps)

esu and negligible nonlinear refraction in very good agreement with the present findings.

The results of calculations of the static molecular (hyper)polarizabilities for **1b**, **2**, and **3** are presented in Table 3.11. Due to the size of the investigated systems, it is computationally not feasible to use post-HF approaches with large basis sets. The MP2 data, obtained with medium-size basis sets, shall serve as a reference for other less accurate methods. Even at the MP2 level of theory we had to restrict the computations to single components of β and γ tensors. What we clearly see is that the influence of electron correlation on β and γ is substantial. In some cases it exceeds even 100 %. For this reason it is interesting to check the performance of density functional theory in determination of molecular hyperpolarizabilities as it accounts for electron correlation effects and yet it is less computationally demanding than post-HF treatments. It has already been shown, however, that traditional functionals have difficulties in prediction of β and γ for organofullerenes [160, 161]. On the other hand, it has been also demonstrated that the long-range corrected functionals, like LC-BLYP [162] or CAM-B3LYP [75], were quite successful in prediction of molecular hyperpolarizabilities of [60]fullerene-chromophore dyads [160, 161] and other organic compounds as well [163–167]. In the case of data for **1b** we see that the CAM-B3LYP values of β and γ, although calculated with relatively small double-ζ basis set, are close to these evaluated at the MP2 level of theory. For 2 the value of $\beta_{zzz}(0)$ calculated using the CAM-B3LYP functional is much larger than the value computed at the MP2 level of theory. The overshoot, however, is not as large as what has been observed for other organofullerenes with the aid of traditional

Table 3.11 The values of molecular (hyper)polarizabilities

	$\beta_{zzz}(0)[au]$	$\gamma_{zzzz}(0)[\times 10^3 au]$	$\gamma_{zzzz}(0)[\times 10^{-36} esu]$
		1B	
HF/3–21G	−1117	132	66
MP2/3–21G	−2006	252	127
CAM-B3LYP/3–21G	−1952	229	115
HF/6–31G(d)	−1206(531)	140(101)	71(51)
MP2/6–31G(d)	−2190	276	139
		2	
HF/6–31G(d)	1(−160)	86(71)	43(36)
MP2/6–31G(d)	−120	188	95
CAM-B3LYP/3–21G	−309	176	89
		3	
HF/3–21G	55	134	67
HF/6–31G	97(338)	187(138)	94(70)
MP2/3–21G	−315	383	193

Averaged values, as defined by Eqs. 3.4 and 3.6, are added in parentheses

functionals [160, 161]. As the reliability of the CAM-B3LYP results occurred to be satisfactory we used this functional for prediction of frequency-dependent second hyperpolarizability for **1b** and **2** (see Fig. 3.23). As mentioned previously, the largest value of second hyperpolarizability has been experimentally observed for **3**. This is consistent with the results of computations presented in this study regardless of the applied level of theory. There is, however, significant discrepancy between the theoretical and experimentally determined second hyperpolarizability (see Tables 3.10 and 3.11), i.e. theory predicts the values of γ to be three orders of magnitude smaller than that determined experimentally. We consider the following possible sources of disagreement: (a) the lack of solvent effects in the theoretical model; (b) neglect of frequency dependence; (c) the presence of cascading effects [168]; (d) picosecond laser pulse duration. Of the two factors which mainly determine the reliability of theoretical predictions, namely (a) and (b), the latter seems to be much more important as far as absolute values of γ are concerned. Indeed, Fig. 3.23 shows that the second hyperpolarizability of **2** determined at 650 nm is three times larger than the static value. One may infer that at experimental wavelength (532 nm) the value would be even larger. It should be also underscored that there is not a perfect match of energy levels observed experimentally and those predicted by theory what further makes the comparison difficult.

3.4.6 Conclusions

In the present work we discussed the linear and nonlinear optical properties of C_{60}-triphenylamine hybrids. In all considered cases, C_{60} serves as an acceptor while triphenylamine unit is chosen as a donor. Following structure-property relationships

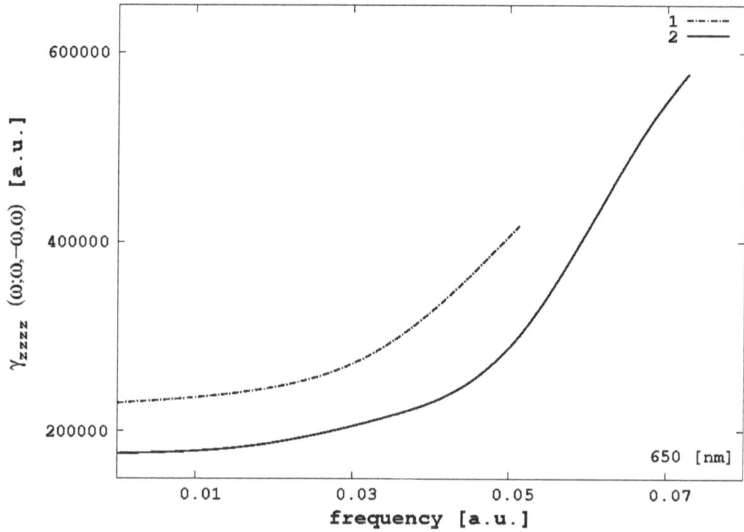

Fig. 3.23 Dependence of the second-order hyperpolarizability on the frequency for **1b** and **2**

established for maximizing two-photon absorption cross-section in organic materials, we considered D-A and D-A-D structural motifs. The molecules were synthesized following the 1,3-dipolar cycloaddition of azomethine ylides onto the skeleton of C_{60} forming the TPhA-based monoadduct, equatorial bis-adduct and dumbell C_{60}. In order to characterize structural and photophysical properties of molecules under study we used NMR, MALDI-TOF-MS, and ATR-IR techniques while the Z-scan technique employing 532 nm, 35 ps laser pulses was used for NLO measurements. A very important finding of the present study, confirmed on purely theoretical basis, is that the total second-order hyperpolarizability of C_{60}-TphA-C_{60} system is several times larger than that of TPhA-C_{60}.

3.4.6.1 Experimental Details

Solvents were purchased from Fluka and Panreac, reagents were purchased from Aldrich and Fluka and C_{60} was purchased from SES Research. Silica gel 60, 70–230 mesh was used in column chromatography. HPLC separations were performed on a LC-9101 instrument using a Cosmosil Buckyprep 20 × 250 mm preparative column and toluene as mobile phase at 10 mL/min. Melting points were measured on a Buchi instrument. NMR spectra were recorded on Bruker AV-500 and 300 spectrometers. MALDI-TOF negative ionization mass spectra were recorded on a Shimadzu spectrometer, using trans-2-[3-(4-tert-butylphenyl)-2-methyl-2-propenylidene]malononitrile as matrix. UV-Vis spectra were recorded on a Perkin Elmer Lambda 19 spectrometer. ATR-IR spectra were recorded on a Bruker Equinox 55 spectrometer.

Bis-fullerene 3. A mixture of C_{60} (60 mg, 0.083 mmol), sarcosine (14 mg, 0.157 mmol) and 4,4'-diformyltriphenylamine 1a (5 mg, 0.017 mmol) in o-dichorobenzene (10 mL) was sonicated for 5 min and then stirred at 130 °C for 1.5 h. After cooling, petroleum ether (7 ml) was added, and the mixture was subjected to column chromatography using 40 % petroleum ether/ o-dichorobenzene to collect unreacted C_{60} (40 mg, 67 %) and then toluene to collect the impure product which was purified by preparative HPLC (retention time 21.83 min), resulting 3 as dark red solid (10 mg, 32 %). ATR-IR (neat): $v(cm^{-1})$ 3026, 2944, 2921, 2849, 2777, 1591, 1505, 1462, 1330, 1314, 1281, 1269, 1178; 1H-NMR (300 MHz, CS_2, insert C6D6): δ(ppm) 3.14 (s, 6H), 4.54 (d, J = 9.3 Hz, 2H), 5.16 (s, 2H), 5.23 (d, J = 9.3 Hz, 2H), 7.15-7.28 (m, 7H), 7.43 (t, J = 7.5 Hz, 2H), 7.88 (bs, 2H); ^{13}C-NMR (63 MHz, CS2, drops of (CD3)2CO) δ(ppm): 40.47, 69.52, 70.49, 78.10, 83.48, 124.02, 124.62, 124.96, 130.10, 130.76, 131.94, 136.40, 136.51, 137.17, 137.31, 139.76, 140.50, 140.70, 142.27, 142.34, 142.66, 142.72, 142.82, 143.20, 143.57, 143.74, 145.00, 145.32, 145.73, 145.87, 145.96, 146.13, 146.39, 146.53, 146.69, 146.78, 146.88, 146.92, 147.14, 147.40, 147.54, 147.82, 148.26, 154.03, 154.24, 154.59, 157.05; MALDI-TOF MS calcd for $C_{144}H_{25}N_3$: 1796, found: m/z 1794.

References

1. Cui, L., Zhao, Y.: Chem. Mater. **16**, 2076 (2004)
2. Haider, J.M., Pikramenou, Z.: Chem. Soc. Rev. **34**, 120 (2005)
3. Sun, X., Liu, Y., Xu, X., Yang, C., Yu, G., Chen, S., Zhao, Z., Qiu, W., Li, Y., Zhu, D.J.: Phys. Chem. B **109**, 10786 (2005)
4. Rahman, G.M.A., Guldi, D.M., Cagnoli, R., Mucci, A., Schenetti, L., Vaccari, L., Prato, M.J.: Am. Chem. Soc. **127**, 10051 (2005)
5. Guo, Z., Du, F., Ren, D., Chen, Y., Zheng, J., Liu, Z., Tian, J.J.: Mater. Chem. **16**, 3021 (2006)
6. Jiang, L., Li, Y.J.: Porphyrins Phtalocyanines **11**, 299 (2007)
7. Kanis, D.R., Ratner, M.A., Marks, T.J.: Chem. Rev. **94**, 195 (1994)
8. Bartkowiak, W.: In: Papadopoulos, M., Sadlej, A.J., Leszczynski, J. (eds.) Non-Linear Optical Properties of Matter. From Molecules to Condensed Phases, Springer, Berlin (2006)
9. Morrall, J.P.L., Humphrey, M.G., Dalton, G.T., Cifuentes, M.P., Samoc, M.: In: Papadopoulos, M., Sadlej, A.J., Leszczynski, J. (eds.) Non-Linear Optical Properties of Matter. From Molecules to Condensed Phases, Springer, Berlin (2006)
10. Coe, B.J.: In: Papadopoulos, M., Sadlej, A.J., Leszczynski, J. (eds.) Non-Linear Optical Properties of Matter. From Molecules to Condensed Phases, Springer, Berlin (2006)
11. Oudar, J.L., Chemla, D.S.J.: Chem. Phys. **66**, 2664 (1977)
12. Hamasaki, R., Ito, M., Lamrani, M., Mitsuishi, M., Miyashita, T., Yamamoto, Y.J.: Mater. Chem. **13**, 21 (2003)
13. Liddell, P.A., Kuciauskas, D., Sumida, J.P., Nash, B., Nguyen, D., Moore, A.L., Moore, T.A., Gust, D.J.: Am. Chem. Soc. **119**, 1400 (1997)
14. Imahori, H., Yamada, K., Hasegawa, M., Taniguchi, S., Okada, T., Sakata, Y.: Angew. Chem. Inter. Ed. Eng. **36**, 2626 (1997)
15. Luo, C., Guldi, D.M., Imahori, H., Tamaki, K., Sakata, Y.J.: Am. Chem. Soc. **122**, 6535 (2000)
16. Imahori, H., Tamaki, K., Guldi, D.M., Luo, C., Fujitsuka, M., Ito, O., Sakata, Y., Fukuzumi, S.J.: Am. Chem. Soc. **123**, 2607 (2001)

17. Imahori, H., Guldi, D.M., Tamaki, K., Yoshida, Y., Luo, C., Sakata, Y., Fukuzumi, S.J.: Am. Chem. Soc. **123**, 6617 (2001)
18. Imahori, H., Tamaki, K., Araki, Y., Sekiguchi, Y., Ito, O., Sakata, Y., Fukuzumi, S.J.: Am. Chem. Soc. **124**, 5165 (2002)
19. Liddell, P.A., Kodis, G., Moore, A.L., Moore, T.A., Gust, D.J.: Am. Chem. Soc. **124**, 7668 (2002)
20. Martin, N., Sánchez, L., Herranz, M.A., Guldi, D.M.J.: Phys. Chem. A **104**, 4648 (2000)
21. Herranz, M.A., Ollescas, B., Martin, N., Luo, C., Guldi, D.M.J.: Org. Chem. **65**, 5728 (2000)
22. Martin, N., Sanchez, L., Guldi, D.M.: Chem. Commun. **113**, (2000)
23. Martin, N., Sanchez, L., Illescas, B., Gonzalez, S., Herranz, M.A., Guldi, D.M.: Carbon **38**, 1577 (2000)
24. Allard, E., Cousseau, J., Oruduna, J., Garin, J., Luo, H., Araki, Y., Ito, O.: Phys. Chem. Chem. Phys. **4**, 5944 (2002)
25. Kreher, D., Hudhomme, P., Gorgues, A., Luo, H., Araki, Y., Ito, O.: Phys. Chem. Chem. Phys. **5**, 4583 (2003)
26. Sanchez, L., Perez, I., Martin, N., Guldi, D.M.: Chem. Eur. J. **9**, 2457 (2003)
27. Guldi, D.M., Maggini, M., Scorrano, G., Prato, M.J.: Am. Chem. Soc. **119**, 974 (1997)
28. D'Souza, F., Zandler, M.E., Smith, P.M., Deviprasad, G.R., Arkady, K., Fujitsuka, M., Ito, O.J.: Phys. Chem. A 106, 649 (2002)
29. Zandler, M.E., Smith, P.M., Fujitsuka, M., Ito, O., D'Souza, F.J.: Org. Chem. **67**, 9122 (2002)
30. Fujitsuka, N., Tsuboya, R., Hamasaki, M., Ito, S., Onodera, S., Ito, O., Yamamoto, Y.J.: Phys. Chem. A 107, 1452 (2003), see http://www.kjemi.uio.no/software/dalton/dalton.html
31. Hogel, H.J.: Phys. Chem. **69**, 755 (1965)
32. Pai, D.M.J.: Chem. Phys. **52**, 2285 (1970)
33. Wang, Y.: Nature **356**, 585 (1992)
34. Wang, Y., Suna, A.J.: Phys. Chem. B **101**, 5627 (1997)
35. Raimundo, J.-M., Blanchard, P., Brisset, H., Akoudad, S., Roncali, J.: Chem. Commun. **939**, (2000)
36. Akhtaruzzaman, M., Tomura, M., Zaman, M.B., Nishida, J., Yamashita, Y.J.: Org. Chem. **67**, 7813 (2002)
37. Edelmann, M.J., Raimundo, J.-M., Utesch, N.F., Diederich, F., Boudon, C., Gisselbrecht, J.-P., Gross, M.: Helv. Chim. Acta **85**, 2195 (2002)
38. Thomas, K.R.J., Lin, M., Velusamy, J.T., Tao, Y.-T., Chuen, C.-H.: Adv. Funct. Mater. **14**, 83 (2004)
39. Sandanayaka, A.S.D., Sasabe, S., Araki, Y., Furosho, Y., Ito, O., Takata, T.J.: Phys. Chem. A **108**, 5145 (2004)
40. Sandanayaka, A.S.D., Taguri, Y., Araki, Y., Ishi-I, T., Mataka, S., Ito, O.J.: Phys. Chem. B **109**, 22502 (2005)
41. Borsenberger, P.M., Fitzgerald, J.J.J.: Phys. Chem. **97**, 4815 (1993)
42. Adachi, Ch., Nagai, K., Tamoto, N.: Appl. Phys. Lett. **66**, 2679 (1995)
43. Shirota, Y.J.: Mater. Chem. **15**, 75 (2005)
44. Sandanayaka, A.S.D., Matsukawa, K., Ishi-I, T., Mataka, S., Araki, Y., Ito, O.J.: Phys. Chem. B **108**, 19995 (2004)
45. Zeng, H.-P., Wang, T., Sandanayaka, A.S.D., Araki, Y., Ito, O.J.: Phys. Chem. A **109**, 4713 (2005)
46. Krieg, T., Petr, A., Barkleit, G., Dunsch, L.: Appl. Phys. Lett. **74**, 3639 (1999)
47. Tagmatarchis, N., Prato, M.: Synlett. **768**, (2003)
48. Tagmatarchis, N., Prato, M.: Struct. Bonding **109**, 1 (2004)
49. Bishop, D.M.: Norman, P. In: Nalwa, H.S. (ed.) Handbook of Advanced Electronic and Photonic Materials and Devices. Academic Press, San Diego (2001)
50. Christiansen, O., Coriani, S., Gauss, J., Hättig, C., Jørgensen, P., Pawłowski, F.: Rizzo, A. In: Papadopoulos, M., Sadlej, A.J., Leszczynski, J. (eds.) Non-Linear Optical Properties of Matter. From Molecules to Condensed Phases, Springer, Berlin (2006)

51. Champagne, B., Perpète, E.A., van Gisbergen, S.J.A., Baerends, E.-J.: Snijders, J.G., Soubra-Ghaoui, C., Robins, K.A., Kirtman, B.J. Chem. Phys. **109**, 10489 (1998)
52. Champagne, B., Perpète, E.A., Jacquemin, D., van Gisbergen, S.J.A., Baerends, E.-J., Soubra-Ghaoui, C., Robins, K.A., Kirtman, B.J.: Phys. Chem. A 104, 4755 (2000)
53. Willets, A., Rice, J.E., Burland, D.M., Shelton, D.P.J.: Chem. Phys. **97**, 7590 (1992)
54. Kurtz, H.A., Stewart, J.J.P., Dieter, K.M.J.: Comp. Chem. **11**, 82 (1990)
55. Quarteroni, A., Sacco, R., Saleri, F.: Méthodes Numériques. Springer Italia, Milano (2007)
56. Frisch, M.J., Trucks, G.W., Schlegel, H.B., Scuseria, G.E., Robb, M.A., Cheeseman; J.R., Montgomery, Jr., J.A., Vreven, T., Kudin, K.N., Burant, J.C., Millam, J.M., Iyengar, S.S., Tomasi, J., Barone, V., Mennucci, B., Cossi, M., Scalmani, G., Rega, N., Petersson, G.A., Nakatsuji, H., Hada, M., Ehara, M., Toyota, K., Fukuda, R., Hasegawa, J., Ishida, M., Nakajima, T., Honda, Y., Kitao, O., Nakai, H., Klene, M., Li, X., Knox, J.E., Hratchian, H.P., Cross, J.B., Bakken, V., Adamo, C., Jaramillo, J., Gomperts, R., Stratmann, R.E., Yazyev, O., Austin, A.J., Cammi, R., Pomelli, C., Ochterski, J.W., Ayala, P.Y., Morokuma, K., Voth, G.A., Salvador, P., Dannenberg, J.J., Zakrzewski, V.G., Dapprich, S., Daniels, A.D., Strain, M.C., Farkas, O., Malick, D.K., Rabuck, A.D., Raghavachari, K., Foresman, J.B., Ortiz, J.V., Cui, Q., Baboul, A.G., Clifford, S., Cioslowski, J., Stefanov, B.B., Liu, G., Liashenko, A., Piskorz, P., Komaromi, I., Martin, R.L., Fox, D.J., Keith, T., Al-Laham, M.A., Peng, C.Y., Nanayakkara, A., Challacombe, M., Gill, P.M.W., Johnson, B., Chen, W., Wong, M.W., Gonzalez, C., Pople, J.A.: Gaussian 03, Rev. D02, Gaussian, Inc., Wallingford (2004)
57. MOPAC2007, J.J.P.: Stewart Computational Chemistry, Version 7.221L web: http://OpenMOPAC.net
58. te Velde, G., Bickelhaupt, F.M., van Gisbergen, S.J.A.: Fonseca Guerra, C., Baerends, E.J., Snijders, J.G., Ziegler, T.J. Comput. Chem. **22**, 931 (2001)
59. Guerra, Fonseca: C., Snijders, J.G., te Velde, G., Baerends, E.J. Theor. Chem. Acc. **99**, 391 (1998)
60. ADF2007.01, SCM, Theoretical Chemistry, Vrije Universiteit, Amsterdam, The Netherlands, http://www.scm.com
61. DALTON, a molecular electronic structure program, Release 2.0 (2005), see http://www.kjemi.uio.no/software/dalton/dalton.html
62. Schmidt, M.W., Baldridge, K.K., Boatz, J.A., Elbert, S.T., Gordon, M.S., Jensen, J.H., Koseki, S., Matsunaga, N., Nguyen, K.A., Su, S., Windus, T.L., Dupuis, M., Montgomery, J.A.J.: Comput. Chem. **14**, 1347 (1993)
63. Luis, J.: M.,; Duran, M.; Champagne, B.; Kirtman, B. J. Chem. Phys. **113**, 5203 (2000)
64. Kirtman, B., Champagne, B., Luis, J.M.J.: Comp. Chem. **21**, 1572 (2000)
65. Bishop, D.M.: Adv. Chem. Phys. **104**, 1 (1998)
66. Monson, P.R., McClain, W.M.J.: Chem. Phys. **53**, 29 (1970)
67. Stewart, J.J.: P J. Comp. Chem. **10**, 209 (1989)
68. Stewart, J.J.: P J. Comp. Chem. **10**, 221 (1989)
69. Greengard, L., Rokhlin, V.J.: Comput. Phys. **73**, 325 (1987)
70. Schmidt, K.E., Lee, M.A.J.: Stat. Phys. **63**, 1223 (1991)
71. Millam, J.M., Scuseria, G.E.: J. Chem. Phys. **106**, 5569 (1997)
72. Hedberg, K., Hedberg, L., Bethune, D., Brown, C., Johnson, R., de Vries, M.: Science **254**, 410 (1991)
73. Loboda, O., Jensen, V.R., Borve, K.J.: Fullerenes. Nanotubes Carbon Nanostruct. **14**, 365 (2006)
74. Foresman, J. B. Æ. Frisch Exploring Chemistry with Electronic Structure Methods, 2nd edn. Gaussian, Inc., Pittsburgh, PA.
75. Sim, F., Chin, S., Dupuis, M., Rice, J.E.J.: Phys. Chem. **97**, 1158 (1993)
76. Zalesny, R., Bartkowiak, W., Toman, P., Leszczynski, J.: Chem. Phys. **337**, 77 (2007)
77. Jacquemin, D., Perpète, E., Scalmani, G., Frisch, M.J., Kobayashi, R., Adamo, C.J.: Chem. Phys. **126**, 144105 (2007)
78. Dreuw, A., Head-Gordon, M.: Chem. Rev. **105**, 4009 (2005)
79. Jacquemin, D., Perpète, E.A., Ciofini, I., Adamo, C.: Chem. Phys. Lett. **405**, 376 (2005)

80. Ciofini, I., Adamo, C., Chermette, H.J.: Chem. Phys. **123**, 121102 (2005)
81. Iikura, H., Tsuneda, T., Yanai, T., Hirao, K.J.: Chem. Phys. **115**, 3540 (2001)
82. Kamiya, M., Sekino, H., Tsuneda, T., Hirao, K.J.: Chem. Phys. **122**, 234111 (2005)
83. Sekino, H., Maeda, Y., Kamiya, M., Hirao, K.J.: Chem. Phys. **126**, 014107 (2007)
84. Yanai, T., Tew, D.P., Handy, N.C.: Chem. Phys. Lett. **393**, 51 (2004)
85. Vignale, G., Kohn, W.: Phys. Rev. Letters **77**, 2037 (1996)
86. Krieger, J.B., Li, Y., Iafrate, G.J.: Phys. Rev. A **45**, 101 (1992)
87. Lipiński, J.: Int. J. Quant. Chem. **34**, 423 (1988)
88. Lipiński, J., Bartkowiak, W.J.: Phys. Chem. A **101**, 2159 (1997)
89. Maggini, M., Scorrano, G.J.: Am. Chem. Soc. **115**, 9798 (1993)
90. Dirk, C.W., Cheng, L.-T., Kuzyk, M.G.: Int. J. Quant. Chem. **43**, 27 (1992)
91. Cronstrand, P., Luo, Y., Ågren, H.: Chem. Phys. Lett. **352**, 262 (2002)
92. Zalesny, R., Bartkowiak, W., Styrcz, S., Leszczynski, J.J.: Phys. Chem. A **106**, 4032 (2002)
93. Xie, Q., Perez-Cordero, E., Echegoyen, L.J.: Am. Chem. Soc. **114**, 3978 (1992)
94. Williams, R.M., Zwier, J.M., Verhoeven, J.W.J.: Am. Chem. Soc. **117**, 4093 (1995)
95. Martín, N., Sánchez, L., Illescas, B., Pérez, I.: Chem. Rev. **98**, 2527 (1998)
96. Cho, M.J., Kim, J.Y., Kim, J.H., Lee, S.H., Dalton, L.R., Choi, D.H.: Bull. Korean Chem. Soc. **26**, 77 (2005)
97. Lee, S.K., Cho, M.J., Kim, G.W., Jun, W.G., Jin, J.-I., Dalton, L.R., Choi, D.H.: Opt. Mater. **29**, 451 (2007)
98. Kato, S.-I., Matsumoto, T., Ishi-i, T., Thiemann, T., Shigeiwa, M., Gorohmaru, H., Maeda, S., Yamashita, Y., Mataka, S.. Chem. Commun. 2342, (2004)
99. Sun, M., Chen, J., Xu, H.J.: Chem. Phys. **128**, 064106 (2008)
100. Göppert-Mayer, M.: Ann. Phys. **9**, 273 (1931)
101. Albota, M., Beljonne, D., Brédas, J.-L., Ehrlich, J.E., Fu, J.-Y., Heikal, A.A., Hess, S.E., Kogej, T., Levin, M.D., Marder, S.R., McCord-Maughon, D., Perry, J.W., Röckel, H.: Rumi, M.: . Science **281**, 1653 (1998)
102. Kogej, T., Beljonne, D., Meyers, F., Perry, J.W., Marder, S.R., Brédas, J.L.: Chem. Phys. Lett. **298**, 1 (1998)
103. Norman, P., Luo, Y., Ågren, H.J.: Chem. Phys. **111**, 7758 (1999)
104. Macak, P., Luo, Y., Norman, P., Ågren, H.J.: Chem. Phys. **113**, 7055 (2000)
105. Lee, W.-H., Lee, H., Kim, J.-A., Choi, J.-H., Cho, M., Jeon, S.-J., Cho, B.R.J.: Am. Chem. Soc. **123**, 10658 (2001)
106. Zhou, X., Ren, A.-M., Feng, J.-K., Liu, X.-J., Zhang, J., Liu, J.: Phys. Chem. Chem. Phys. **4**, 4346 (2002)
107. Liu, X.-J., Feng, J.-K., Ren, A.-M., Zhou, X.: Chem. Phys. Lett. **373**, 197 (2003)
108. Bartkowiak, W., Zalesny, R., Leszczynski, J.: Chem. Phys. **287**, 103 (2003)
109. Masunov, A., Tretiak, S.J.: Phys. Chem. B **108**, 899 (2004)
110. Ehrlich, J.E., Wu, X.L., Lee, I.Y.S., Hu, Z.Y., Röckel, H., Marder, S.R., Perry, J.W.: Opt. Lett. **1997**, 22 (1843)
111. Parthenopoulos, D.M., Rentzepis, P.M.: Science **245**, 843 (1989)
112. Li, X.-D., Cheng, W.-D., Wu, D.-S., Lan, Y.-Z., Zhang, H., Gong, Y.-J., Li, F.-F., Shen, J.J.: Phys. Chem. B **109**, 5574 (2005)
113. Padmawar, P.A., Rogers, J.E., He, G.S., Chiang, L.Y., Tan, L.-S., Canteenwala, T., Zheng, Q., Slagle, J.E., McLean, D.G., Fleitz, P.A., Prasad, P.N.: Chem. Mater. **18**, 4065 (2006)
114. Xie, R., Bryant, G.W., Jensin, L., Zhao, J., Smith Jr, V.H.J.: Chem. Phys **118**, 8621 (2003)
115. Jonsson, D., Norman, P., Ruud, K., Ågren, H.J.: Chem. Phys. **109**, 572 (1998)
116. Jansik, B., Salek, P., Jonsson, D., Vahtras, O., Ågren, H.J.: Chem. Phys. **122**, 054107 (2005)
117. Ecklund, P.: Bull. Am. Phys. Soc. **37**, 191 (1992)
118. Hebard, A.F., Haddon, R.C., Fleming, R.M., Korton, A.R.: Appl. Phys. Lett. **59**, 2109 (1991)
119. Lambin, P.H., Lucas, A.A., Vingneron, J.P.: Phys. Rev. B **46**, 1794 (1992)
120. Meth, J.S., Vanherzeele, H., Wang, Y.: Chem. Phys. Lett. **197**, 26 (1992)
121. Antoine, R., Dugourd, P., Rayane, D., Benichou, E., Broyer, M., Chandezon, F., Guet, C.J.: Phys. Chem. **110**, 9771 (1999)

122. Ballard, A., Bonin, K., Louderback, J.: J Phys. Chem. **113**, 5732 (2000)
123. Norman, P., Luo, Y., Jonsson, D., Ågren, H.J.: Chem. Phys. **106**, 8788 (1997)
124. Peeters, E., van Hal, P., Knol, J., Brabec, C.J., Sariciftci, N.S., Hummelen, J.C., Janssen, R.A.J.: J. Phys. Chem. B **104**, 10174 (2000)
125. D'Souza, F., Gadde, S., Schumacher, A.L., Zandler, M.E., Sandanayaka, A.S.D., Araki, Y., Ito, O.: J. Phys. Chem. C **111**, 11123 (2007)
126. Fukuda, T., Masuda, S., Kobayashi, N.: J. Am. Chem. Soc. **129**, 5472 (2007)
127. Kaunisto, K., Chukharev, V., Tkachenko, N.V., Lemmetyinen, H.: Chem. Phys. Lett. **460**, 241 (2008)
128. Koudoumas, E., Konstantaki, M., Mavromanolakis, A., Couris, S., Fanti, M., Zerbetto, F., Kordatos, K., Prato, M.: Chem. Eur. J. **9**, 1529 (2003)
129. Tanihara, J., Ogawa, K., Kobuke, Y.: J. Photochem. Photobiol. A **178**, 140 (2006)
130. Xenogiannopoulou, E., Medved, M., Iliopoulos, K., Couris, S., Papadopoulos, M.G., Bonifazi, D., Sooambar, C., Mateo-Alonso, A., Prato, M.: Chem. Phys. Chem. **2007**, 8 (1056)
131. Rae, M., Perez-Balderas, F., Baleizao, C., Fedorov, A., Cavaleiro, J.A.S., Tome, A.C., Berberan-Santos, M.N.: J. Phys. Chem. B **110**, 12809 (2006)
132. Brites, M.J., Santos, C., Nascimento, S., Gigante, B., Luftmann, H., Fedorov, A., Berberan-Santos, M.N.: New J. Chem. **2006**, 30 (1036)
133. Petsalakis, I.D., Tagmatarchis, N., Theodorakopoulos, G.: J. Phys. Chem. C **111**, 14139 (2007)
134. Sheka, E.F.: Int. J. Quant. Chem. **107**, 2361 (2007)
135. Toivonen, T.L.J., Hukka, T.I.: J. Phys. Chem. A **111**, 4821 (2007)
136. Loboda, O.: In: Quantum Modeling of Spin-Selective Processes of the Triplet State VDM, Saarbrucken (2009)
137. Zeng, H.-P., Wang, T., Sandanayaka, A.S.D., Araki, Y., Ito, O.: J. Phys. Chem. A **109**, 4713 (2005)
138. Sandanayaka, A.S.D., Taguri, Y., Araki, Y., Ishi-I, T., Mataka, S., Ito, O.: J. Phys. Chem. B **109**, 22502 (2005)
139. Shirota, Y.: J. Mater. Chem. **15**, 75 (2005)
140. Bell, T.D.M., Stefan, A., Masuo, S., Vosch, T., Lor, M., Cotlet, M., Hofkens, J., Bernhardt, S., Mullen, K., van der Auweraer, M., Verhoeven, J.W., De Schryver, F.C.: Chem. Phys. Chem. **6**, 942 (2005)
141. Couris, S., Koudoumas, E., Ruth, A.A., Leach, S.: J. Phys. B: At. Mol. Opt. Phys. **2**, 4537 (1995)
142. Guldi, D.M.: Chem. Commun. **321**, (2000)
143. Sanchez, L., Herranz, M.A., Martin, N.: J. Mater. Chem. **15**, 1409 (2005)
144. Rotas, G., Tagmatarchis, N.: Tetrahedron Lett. **50**, 398 (2009)
145. Mallegol, T., Gmouth, S., Meziane, M.A.A., Blanchard-Desce, M.O.: Mongin, Synthesis 1771 (2005)
146. El-Khouly, M.E., Kim, J.H., Kwak, M., Choi, C.S., Ito, O., Kay, J.-H.: Bull. Chem. Soc. Jpn. **80**, 2465 (2007)
147. Kurtz, H.A., Stewart, J.J.P., Dieter, K.M.: J. Comp. Chem. **11**, 82 (1990)
148. Frisch, M.J., Trucks, G.W., Schlegel, H.B., Scuseria, G.E., Robb, M.A., Cheeseman, J.R., Montgomery, J.A., Vreven, T., Kudin, K.N., Burant, J.C., Millam, J.M., Iyengar, S.S., Tomasi, J., Barone, V., Mennucci, B., Cossi, M., Scalmani, G., Rega, N., Petersson, G.A., Nakatsuji, H., Hada, M., Ehara, M., Toyota, K., Fukuda, R., Hasegawa, J., Ishida, M., Nakajima, T., Honda, Y., Kitao, O., Nakai, H., Klene, M., Li, X., Knox, J.E., Hratchian, H.P., Cross, J.B., Bakken, V., Adamo, C., Jaramillo, J., Gomperts, R., Stratmann, R.E., Yazyev, O., Austin, A.J., Cammi, R., Pomelli, C., Ochterski, J.W., Ayala, P.Y., Morokuma, K., Voth, G.A., Salvador, P., Dannenberg, J.J., Zakrzewski, V.G., Dapprich, S., Daniels, A.D., Strain, M.C., Farkas, O., Malick, D.K., Rabuck, A.D., Raghavachari, K., Foresman, J.B., Ortiz, J.V., Cui, Q., Baboul, A.G., Clifford, S., Cioslowski, J., Stefanov, B.B., Liu, G., Liashenko, A., Piskorz, P., Komaromi, I., Martin, R.L., Fox, D.J., Keith, T., Al-Laham, M.A., Peng, C.Y., Nanayakkara, A., Challacombe, M., Gill, P.M.W., Johnson, B., Chen, W., Wong, M.W., Gonzalez, C.J.A.: Gaussian 03, ReV. E01, Gaussian, Inc.: Wallingford, CT (2004)

149. Schmidt, M.W., Baldridge, K.K., Boatz, J.A., Elbert, S.T., Gordon, M.S.: Jensen, J.H., Koseki, S., Matsunaga, N., Nguyen, K.A., Su, S., Windus, T.L., Dupuis, M., Montgomery, J.A. J. Comput. Chem. **14**, 1347 (1993)
150. DALTON: a molecular electronic structure program, Release 2.0 (2005)
151. Sheik-Bahae, M., Said, A., Wei, T., Hagan, D., Stryland, E.V.: IEEE J. Quantum Electron. **26**, 760 (1990)
152. Kordatos, K., Da Ros, T., Prato, M., Bensasson, R.V., Leach, S.: Chem. Phys. **293**, 263 (2003)
153. Perpete, E.A., Wathelet, V., Preat, J., Lambert, C., Jacquemin, D.: J. Chem. Theory Comput. **2**, 434 (2006)
154. Jacquemin, D., Bouhy, D., Perpete, E.A.: J. Chem. Phys. **124**, 204321 (2006)
155. Maggini, M., Scorrano, G.: J. Am. Chem. Soc. **115**, 9798 (1993)
156. Iliopoulos, K., Couris, S., Hartnagel, U., Hirsch, A.: Chem. Phys. Lett. **448**, 243 (2007)
157. Iliopoulos, K.: In: Investigation of the Nonlinear Optical Response of Fullerene Derivatives and Nanoparticles for Optical Sensing Applications, PhD thesis, University of Patras, Greece (2008)
158. Albota, M., Beljonne, D., Bredas, J.-L., Ehrlich, J.E., Fu, J.-Y., Heikal, A.A., Hess, S.E., Kogej, T., Levin, M.D., Marder, S.R., McCord-Maughon, D., Perry, J.W.: Rockel, H., Rumi, M., Subramaniam, G., Webb, W.W., Wu, X.-L., Xu, C. Science. **281**, 1653 (1998)
159. Mavritsky, O.B., Egorov, A.N., Petrovsky, A.N., Yakubovsky, K.V., Blau, W., Weldon, D.N., Henari, F.Z.: Proc. SPIE **2854**, 254 (1996)
160. Loboda, O., Zalesny, R., Avramopoulos, A., Papadopoulos, M., Artacho, E.: AIP Conf. Proc. **1108**, 198 (2009)
161. Loboda, O., Zalesny, R., Avramopoulos, A., Luis, J.M., Kirtman, B., Tagmatarchis, N., Reis, H., Papadopoulos, M.: J. Phys. Chem. A **113**, 1159 (2009)
162. Iikura, H., Tsuneda, T., Yanai, T., Hirao, K.: J. Chem. Phys. **115**, 3540 (2001)
163. Jacquemin, D., te PerpÃ, E.A., Scalmani, G., Frisch, M.J., Kobayashi, R., Adamo, C.: J. Chem. Phys. **126**, 144105 (2007)
164. Kirtman, B., Bonness, S., Ramirez-Solis, A., Champagne, B., Matsumoto, H., Sekino, H.: J. Chem. Phys. **128**, 114108 (2008)
165. Krawczyk, P., Kaczmarek, A., Zalesny, R., Matczyszyn, K., Bartkowiak, W., Ziolkowski, M., Cysewski, P.: J. Mol. Model. **15**, 581 (2009)
166. Limacher, P.A., Mikkelsen, K.V.: LÃ$\frac{1}{4}$thi, H.P. J. Chem. Phys. **130**, 194114 (2009)
167. Zalesny, R., Wojcik, G., Mossakowska, I., Bartkowiak, W., Avramopoulos, A., Papadopoulos, M.G.: J. Mol. Struct. (THEOCHEM) 907, 46 (2009)
168. Meredith, G.R.: Chem. Phys. Lett. **92**, 165 (1982)

Supplementary material

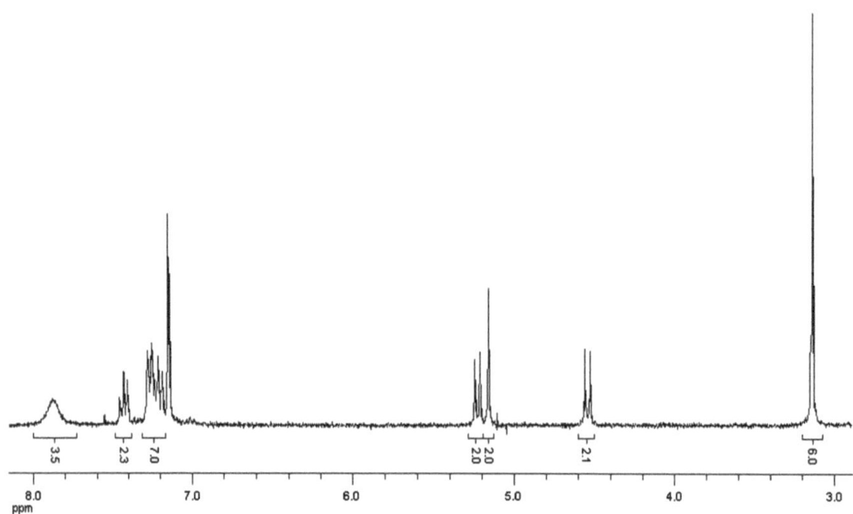

Fig. S1

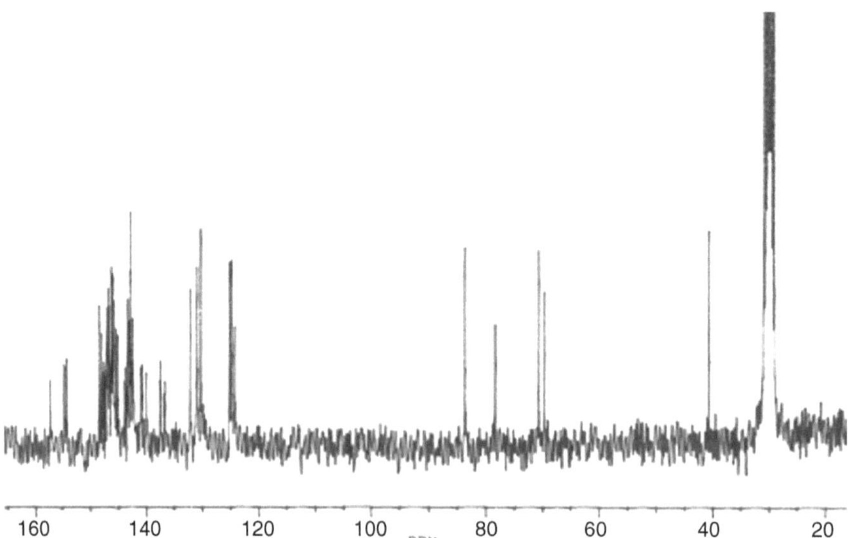

Fig. S2

gr108 486 - Signal 2
2500
2000
1500
1000
500
0
mV
* 21,83
* 11,18
10
20
30
40
Minutes

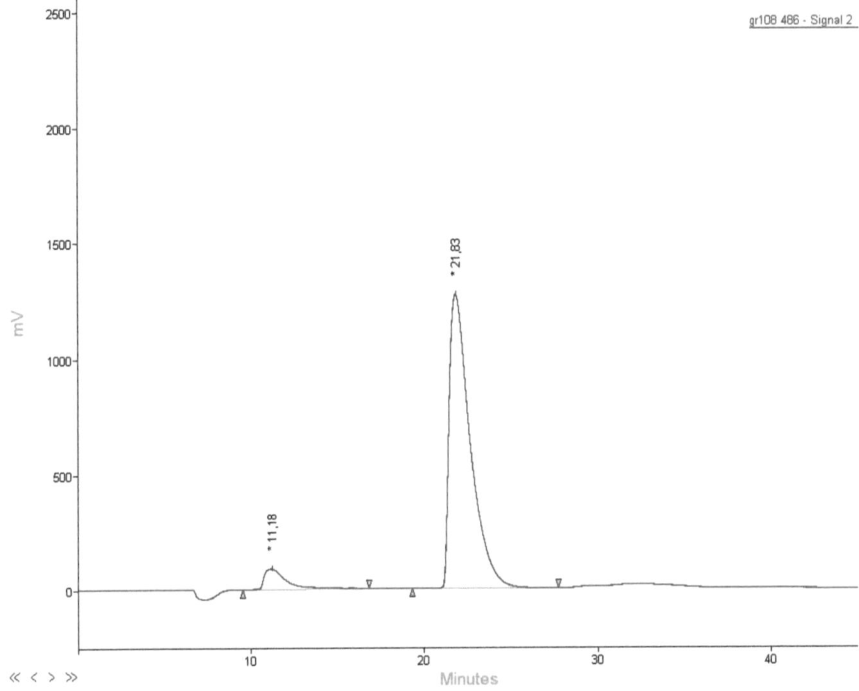

Fig. S3

Chapter 4
Endohedral Metallofullerenes

Since their discovery endohedral fullerenes have been extensively investigated because of their novel structure and properties [1]. Electrical properties have been of major interest owing to a variety of possible applications ranging from qubits for quantum computation [2] to organic photovoltaic devices [3]. Our interest in this connection lies in the theoretical determination of the linear and nonlinear optical properties, i.e. the (hyper)polarizabilities, of these materials. For that purpose we have chosen initially to study the prototypical metal endohedral fullerene, $Li@C_{60}$, and its cation $[Li@C_{60}]^+$.

It has been recognized for some time now [4] that large amplitude vibrational motions, due to weak interactions between the dopant Li atom and the fullerene cage, could give rise to large vibrational (hyper)polarizabilities. Thus, both the pure electronic and vibrational contributions need to be examined. So far both contributions have been treated only at a rudimentary level, to a large extent because of the large size of the fullerene cage and the open shell character of the neutral. Furthermore, only static properties have been considered as far as computations are concerned. The pure electronic first hyperpolarizability [5] and second hyperpolarizability [6] of the neutral have been calculated using an uncoupled Hartree-Fock scheme with molecular orbitals obtained from a restricted open-shell Hartree-Fock calculation. These papers also contain experimental measurements. Very recently Yaghobi et al. [7, 8] used a modified Su-Schrieffer-Heeger (Huckel-type) Hamiltonian [9], coupled with a sum-over-states procedure, to obtain the $Li@C_{60}$ electronic linear polarizability and second hyperpolarizability tensors.

Whereas all the theoretical papers mentioned here discuss the vibrational contribution to the (hyper)polarizabilities, only Whitehouse and Buckingham [4] made an attempt to calculate these properties for the type of system in which we are interested. They used a much simplified potential, in conjunction with a classical analysis—estimated to be valid above 20 K—to obtain a (temperature-dependent) expression for the vibrationally averaged dipole moment. Then, from the field-dependence of this expression, formulas for the vibrational linear polarizability and second hyperpolarizability were extracted and the former quantity was evaluated for the $[Li@C_{60}]^+$

O. Loboda, *Quantum-Chemical Studies on Porphyrins, Fullerenes and Carbon Nanostructures*, Carbon Nanostructures, DOI: 10.1007/978-3-642-31845-0_4,

cation. Their work indicated that the vibrational contribution could be many times larger than the electronic contribution for this system.

As concerns endohedral fullerenes in general, some calculations of their polarizabilities have been reported, usually employing DFT methods [10–15], but reports on vibrational polarizabilities are rare. We mention here the work of Pederson et al. [16], who computed the vibrational polarizability of Kr@C_{60}, as well as of C_{60} itself, in the lowest-order of perturbation theory (double-harmonic approximation). At that level, these contributions were found to be very small compared with the electronic property.

There have been many important advances in computational capability and theoretical methodolgy since the articles on Li doped C_{60} noted above have appeared. While these articles establish a clear interest in the linear and nonlinear optical properties of the prototypical neutral and cation, the time is ripe for a significantly improved treatment. That is the goal of our current work.

4.1 Methods of Investigation

The geometry of Li@C_{60} and the singly charged cation were optimized at the DFT level using the B3LYP functional in the unrestricted version for the former and the restricted version for the latter. We note here that spin contamination was not an issue in the unrestricted DFT calculations, neither with nor without applied electric fields. In all cases, the expectation value of the total spin squared operator, $\langle S^2 \rangle$ for the converged Li@C_{60} wavefunctions was found to be between 0.755 and 0.756, which is close to the exact value of 0.75 for a pure doublet. Due to the large size of the molecule, as well as the tight convergence requirement for calculating vibrational (hyper)polarizabilities, the rather small 6-31G basis set was employed in the latter calculations (some 6-31+G values are included for comparison). For the electronic properties, the 6-31G basis was found to be inadequate, thus they were additionally computed with the 6-31+G and 6-31+G* basis sets. In addition, several geometry optimisations were done with 6-31+G and 6-311G*, although no Hessians could be computed with these large basis sets.

The static electronic polarizabilites ($\alpha^e_{\alpha\beta} \equiv \alpha^e_{\alpha\beta}(0;0)$), first hyperpolarizabilities ($\beta^e_{\alpha\beta\gamma} \equiv \beta^e_{\alpha\beta\gamma}(0;0,0)$), and second hyperpolarizabilities ($\gamma^e_{\alpha\beta\gamma\delta} \equiv \gamma^e_{\alpha\beta\gamma\delta}(0;0,0,0)$) are defined by the Taylor series expansions for the dipole moments $\mu_\alpha(F)$ [17] or energies $E(F)$ [17], in terms of the static field F:

$$E(F) = E(0) - \mu^e_i F_i - \frac{1}{2}\alpha^e_{ij} F_i F_j - \frac{1}{6}\beta^e_{ijk} F_i F_j F_k - \frac{1}{24}\gamma^e_{ijkl} F_i F_j F_k F_l - \cdots \tag{4.1}$$

$$\mu_i(F) = -\frac{\partial E(F)}{\partial F_i} = \mu^e_i + \alpha^e_{ij} F_j + \frac{1}{2}\beta^e_{ijk} F_j F_k + \frac{1}{6}\gamma^e_{ijkl} F_j F_k F_l + \cdots \tag{4.2}$$

From these computed properties, i.e. the dipole moment and energy, the (hyper)polarizabilites were obtained using the Romberg differentiation procedure. Convergence difficulties in the DFT self-consistent orbital calculations severely limit the accuracy of the numerical differentiation as well as the range of fields that can be used. The problem increases for larger fields and thereby impacts the accuracy of the (hyper)polariabilities. Thus, the second hyperpolarizability could be obtained with sufficient statistical confidence from the dipole moments, but not from the energies. Generally, a third-order Romberg differentiation [18] with a minimal applied field strength of 0.001 au was applied. In some cases, only a second-order Romberg differentiation was possible due to convergence problems, but in some other cases even a higher-order was used to ensure the reliability of the values. In addition to the DFT computations, selected (Hartree-Fock) HF and second-order Møller-Plesset (MP2) computations were performed with the 6-31G basis set to provide a comparison with traditional wavefunction theory methods. The computations were done with Gaussian03 [19] and Gaussian09 [20].

Some of the problems mentioned above could have been avoided by using an analytical derivative method, e.g. analytical response theory, as implemented for linear, quadratic and cubic response functions in time-dependent DFT by Jansik et al. [21] in the program package Dalton [22]. This method has also been extended using spin-restricted DFT for open-shell systems [23, 24]. However, trials to compute the (hyper)polarizabilities of $Li@C_{60}$ using the smallest basis set (6-31G) were successful only for linear polarizabilities; convergence problems in determining the response vectors prevented calculation of the hyperpolarizabilities. Considering that these calculations were quite expensive, and finite fields were needed for the nuclear relaxation treatment of vibrational contributions, we decided to employ finite field techniques throughout. We mention, finally, that the polarizabilities obtained from spin-restricted analytical response theory were nearly identical to those calculated by unrestricted finite field methods, showing again that spin-contamination is not a problem in our case.

Although there has been a lot of progress in the last few years in the field of computing vibrational (hyper)polarizabilities based on vibrational self-consistent field theory and correlated versions thereof (see e.g. Refs. [25, 26]), the corresponding methods are computationally still much to expensive for the large systems of interest here. Thus, the vibrational contributions were mostly computed using the finite field approach pioneered by Bishop, et al. [27], and later implemented by Luis et al. [28]. In this approach, the molecular geometry is first optimized in the presence of a static electric field while strictly maintaining the Eckart conditions [28]. Then the difference in the static electronic properties due to the change in geometry induced by the field is expanded as a power series in the field. Each term in the expansion yields the sum of a static electronic (hyper)polarizability plus a nuclear relaxation (NR) vibrational term. For example, the change of the dipole moment and of the linear polarizability are given by [27]:

$$\Delta\mu_i(F, R_F) = a_{1,ij} F_j + \frac{1}{2} b_{1,ijk} F_j F_k + \frac{1}{6} g_{1,ijkl} F_j F_k F_l + \cdots \quad (4.3)$$

$$\Delta\alpha_{ij}(F, R_F) = b_{2,ijk} F_k + \cdots \quad (4.4)$$

with

$$a_{1,ij} = \alpha^e_{ij}(0; 0) + \alpha^{nr}_{ij}(0; 0) \quad (4.5)$$

$$b_{1,ijk} = \beta^e_{ijk}(0; 0, 0) + \beta^{nr}_{ijk}(0; 0, 0) \quad (4.6)$$

$$g_{1,ijkl} = \gamma^e_{ijkl}(0; 0, 0, 0) + \gamma^{nr}_{ijkl}(0; 0, 0, 0) \quad (4.7)$$

$$b_{2,ijk} = \beta^e_{ijk}(0; 0, 0) + \beta^{nr}_{ijk}(-\omega; \omega, 0)_{\omega \to \infty} \quad (4.8)$$

The argument R_F implies structure relaxation in the field, and P^{nr} means the nuclear relaxation part of P, while the subscript $\omega \to \infty$ invokes the so-called "infinite optical frequency (IOF)" approximation. In principle, this procedure allows one to obtain most of the major dynamic vibrational NR contributions in addition to the purely static ones of Eqs. 4.5–4.7. The linear term in the electric field expansion of Eq. (4) gives the dc-Pockels effect; the quadratic term gives the optical Kerr Effect; and the linear term in the expansion of beta yields dc-second harmonic generation (all in the IOF approximation). For laser frequencies in the optical region it has been demonstrated that the latter approximation is normally quite accurate [29–31]. In fact, this approximation is equivalent to neglecting terms of the order $(\omega_v/\omega)^2$ with respect to unity (ω_v is a vibrational frequency). In terms of Bishop and Kirtman perturbation theory [32–34] all vibrational contributions through first-order in mechanical and/or electrical anharmonicity, and some of second-order, are included in the NR treatment [35].

The remaining (higher-order) vibrational contributions can, in principle be computed as well using a related formulation [36]. However, that treatment requires computation of the field-dependent zero-point vibrationally averaged properties, which was not feasible for the systems studied here because of their large size and complicated potential energy surface (PES). Indeed, of the dynamic properties mentioned above, we were only able to obtain the dc-Pockels Effect due to instabilities for high fields that will be described later.

Generally, field strengths from 0.0001 au up to 0.0128 au were tried in the Eckart-constrained optimizations and the energies and dipole moments of the successfully optimized structures were subjected to a numerical Romberg differentiation. As in the case of the electronic properties, the numerical differentiation of the energies was too unstable to yield all the properties of interest. However, the numerical differentiation of field-induced dipole moments allowed us to obtain stable values for most of the components of the NR contribution to the static α , β and γ and to the IOF approximation for the dc-Pockels first hyperpolarizability. Due to the high computational cost of these calculations, the 6-31G basis set had to be used. A few control calculations with the 6-31+G basis set showed that the influence of diffuse basis functions on the vibrational properties is not negligible, but smaller than on the electronic properties.

4.2 Results and Discussion

4.2.1 Geometry Optimization

Zhang et al. [37] have calculated the UB3LYP/6-311G* potential energy surface (PES) for motion of Li along five different rays passing through the center of an undistorted fullerene cage in $Li@C_{60}$. The two most important rays, as far as the structure is concerned, were along the line from the cage center to the center of a C_6 hexagon (symmetry C_{3v}) and along the line from the cage center to the center of a C_5 pentagon (symmetry C_{5v}). Since localization due to cage distortion can be important, as they found, we carried out geometry optimizations for near C_{3v} and near C_{5v} symmetry at the same level while allowing the cage to fully relax. We also obtained the stationary point at the near icosahedral symmetry under the same conditions. In all cases, it was necessary to lower the actual symmetry to obtain the optimized structure, due to SCF convergence problems in the high symmetry calculations. Zhang et al. [37] report that similar difficulties occurred in their calculations.

In the case of approximate I_h symmetry the cage was slightly distorted yielding a C_s optimized structure with the Li atom slightly (0.015 Å for UB3LYP/6-31G) removed from the cage center. As already well-known this stationary point is not a minimum; in fact, there are four imaginary frequencies. For the two minima (near C_{5v} and near C_{3v}) the Li atom shifts about 0.1 Å off the ray that goes from the center of the cage to the center of the polygon and the symmetry is again reduced to C_s. In both instances, the Li atom was located at about 1.5 Å from the center of the cage in the optimized structure. The eccentric position of Li in $Li@C_{60}$ has been interpreted in terms of dispersion and repulsion [38, 39] interactions.

All of the above results agree semi-quantitatively with Zhang et al. [37] as expected (see further below). The structure with approximate C_{3v} symmetry was found to be 3.4 kJ/mol (1.6 kJ/mol) more stable than the one with C_{5v} symmetry using UB3LYP/6-311G* (UB3LYP/6-31G). Our value is slightly higher than the one found by Zhang et al. (2.6 kJ/mol). The reason for this difference may be due to cage relaxation and/or small deviations of the Li atom from the fixed ray they employed. It is also possible that the two minima are not directly related; as shown further below, there seem to be several minima close by. The energy difference between the near-Ih and near C_{3v} symmetry structures was found to be 56.5 kJ/mol at the 6-31G/UB3LYP level. Since we are interested in the ground vibrational state all further investigations were focused on the most stable near-C_{3v} structure. The optimization using the 6-31+G basis was started from the 6-31G optimized structure, and the final optimized structure was very close to the starting one, as expected.

The geometry of the monovalent cation $[Li@C_{60}]^+$ was also determined for the near C_{3v} symmetry, using restricted B3LYP and the 6-31G basis set. A control optimization using B3LYP/6-311G* did not show any substantial structural differences. At the minimum, the Li atom is about 1.4 Å off the center of mass of the cage. The average C–C bond-length (1.4406 Å) is nearly the same as that of the neutral (1.4412 Å). However, the cation is somewhat more spherical. As a measure of

Table 4.1 Electronic contribution to dipole moment (μ), first- (α^e), second- (β^e) and third-order polarizability (γ^e) for Li@C_{60} and [Li@C_{60}]$^+$ using different levels of theory and different basis sets, for the (U)B3LYP/6-31G optimized geometry

		Li@C_{60}				[Li@C_{60}]$^+$		
Method	Basis set	μ_z^e	α_{zz}^e	β_{zzz}^e	$\gamma_{zzzz}^e \times 10^3$	α_{zz}^e	β_{zzz}^e	$\gamma_{zzzz}^e \times 10^3$
(U/R)B3LYP	6-31G	0.049	508.0	1540.8	−39	468.9	−237	28
(U/R)B3LYP	6-31+G	0.328	578.56	1839	64	520.2	−53	99
(U/R)B3LYP	6-31+G*	0.390	589.03	1532.6	66	533.9	−118	99
(U/R)HF	6-31G	−0.18	479.0	5540	900	445.8	−83	54
(U/R)MP2	6-31G	1.39	527.1	1362.6	–	463.6	−192	–

All values in a.u.

the sphericality we use $\Delta I = [(I_a - I_b)^2 + (I_a - I_c)^2 + (I_b - I_c)^2]^{1/2}]$, where I_x is the principal component of the cage inertia tensor, with respect to the center of mass, in the x-direction. Our values are $\Delta I = 1.4\,\mathrm{g\,Å^2/mol}$ for the cation and $\Delta I = 31.5\,\mathrm{g\,Å^2/mol}$ for the neutral. For comparison, the average moment of inertia $I = 1/3(I_a + I_b + I_c)$ is about 3050 g Å^2/mol for both species. Finally, the coordinates of both near C_{3v} optimized structures can be obtained from the authors.

4.2.2 *Electronic Properties*

In order to assess the reliability of the level of theory chosen, we show in Table 4.1 the computed static electronic properties of Li@C_{60} and [Li@C_{60}]$^+$, along the dipole moment direction (defined as z), at different levels of theory and with different basis sets. Due to the convergence problems mentioned above, it was not possible to determine the second hyperpolarizability, γ, at the (U)MP2 level for either species. Because of the horizontal shift of the Li atom in the optimization process (*vide supra*), the z-axis does not coincide with the axis containing the cage center of mass and the Li atom, but is tilted away by about 10°. For [Li@C_{60}]$^+$ the dipole moment is determined by placing the (arbitrary) origin at the cage center of mass.

The rather small dipole moment of the neutral depends strongly on the basis set and correlation treatment. Fortunately our interest lies in the (hyper)polarizabilities. Nonetheless, we can say that our dipole moment results are in qualitative agreement with the value computed by Campbell et al. [5], but much smaller than reported by Li and Tomanek [40]. The comparison of our calculated (hyper)polarizabilities with those of Campbell et al. will be made later. We note that the addition of diffuse functions to the 6-31G basis is always crucial. But whereas the further addition of polarization functions has a minor effect on α and γ, for β they partially (or totally) offset the effect of the diffuse functions. Comparison of HF and (U)MP2 shows that correlation has a very large effect on β, but not α The fact that (U)B3LYP yields values similar to (U)MP2 suggests that both account fairly well for correlation (even though the calculations are only at the 6-31G level). Due to the unavailability of γ at

Table 4.2 Electronic (e) contribution to diagonal components of α, β and γ for Li@C_{60}, [Li@C_{60}]$^+$, C_{60}, C_{60}^- and Li calculated at the UB3LYP/6-31+G* level of theory

$i =$	Li@C_{60}^{a}			[Li@C_{60}]$^{+b}$			
	x	y	z	x	y	z	
α^e_{ii}	560.03	590.29	589.03	533.86	533.86	533.89	
β^e_{iii}	0.0	−290.0	1532.6	0	0	−118.1	
$\gamma^e_{iiii} \times 10^3$	293	−20	66	102	102	99	
$i, j =$	x, y	x, z	y, z	z, y	x, y	x, z	y, z
β^e_{iij}	98	−1104	514	−58	0	−52	−52
$i =$	C_{60}^{-} [a]			C_{60} [b]			Li
	x	y	z	x	y	z	$x = y = z$
α^e_{ii}	575.9	617.7	620.5	550.5	550.5	551.3	138.6, 143[c], 142.9[d]
β^e_{iii}	0	−88	441	0	0	−11	0
$\gamma^e_{iiii} \times 10^3$	211	−46	−86	136	135	136	−250, 568[b], 631[c]

A couple of larger basis set calculations are reported for Li
[a] At the geometry of Li@C_{60}/UB3LYP/6-31G
[b] At the geometry of [Li@C_{60}]$^+$/B3LYP/6-31G
[c] aug-cc-pVQZ basis
[d] aug-cc-pV5Z basis

the (U)MP2 level, no conclusions can be drawn in this respect for the second hyperpolarizability. The cation properties are somewhat less sensitive to correlation than those of the neutral. Overall, we conclude that (U)B3LYP/6-31+G* is the minimal level required to obtain reliable electronic properties. Finally, the large change in hyperpolarizabilities upon going from the cation to the neutral is not unexpected in view of the additional electron in a (formerly) unoccupied orbital localized on the C_{60} moiety. For γ the effect is more evident for the other two diagonal components shown in Table 4.2 (see below).

In Table 4.2 we show the computed static (U)B3LYP/6-31+G* electronic properties of Li@C_{60}, [Li@C_{60}]$^+$, C_{60}^-, C_{60} and Li. The several additional species were included for comparison of Li@C_{60} with both non-interacting Li + C_{60} and Li^+ + C_{60}^-, as well as for comparison of [Li@C_{60}]$^+$ with non-interacting Li^+ + C_{60}. For α and γ only selected components useful for this purpose are displayed whereas, for β, all symmetry allowed components are given (see later). The 6-31+G* basis is inadequate for the Li atom. Although it is not important here, we have added results for Li atom obtained with two larger basis sets (aug-cc-pVQZ and aug-cc-pV5Z) showing that the negative 6-31+G* value of γ becomes positive for the larger Dunning basis sets. The properties of Li^+ are negligible [41] in the current context and are not taken into account in the discussion below. For our comparisons the geometry of the endohedral fullerenes was optimized at the (U)B3LYP/6-31G level and the same geometry was retained for the non-interacting species. In principle, a BSSE correction should be applied to the Li-doped fullerenes, but is omitted since it would have no effect on our conclusions.

As compared to the hypothetical non-interacting species the interaction between the Li^+ cation and either the C_{60}^- or C_{60} cage leads to a moderate reduction of the diagonal polarizabilities and, in the case of $[LiC_{60}]^+$ also of the second hyperpolarizabilities. The reduction for α may be due to a contraction of the electron density caused by the attraction of the cation. Such an explanation will not suffice for γ since the effect of the interaction on the diagonal components is quite different for Li@C_{60}, and the second hyperpolarizability is not simply related to the size of the electron distribution. The first hyperpolarizabilities arise from asymmetry of the charge distribution and are, consequently, strongly enhanced in the endohedral species.

Campbell et al. [6] used an uncoupled approximation to coupled-perturbed HF theory—or, as they prefer to call it, a "computationally expensive tight-binding approach" to compute the hyperpolarizabilities of Li@C_{60}, using the 6-31G* basis set, and obtained for γ the values $\sim$(320, 540, -320) $\times$ 10^3 au, for the x, y, and z diagonal components respectively. While the x value is quite close to ours, the other two values do not agree even in sign. In Ref. [5], Campbell et al. also computed the first hyperpolarizability using the same methodology. The values they obtained ($\beta_{zzz} \sim 15000$ au, $\beta_{yyy} \sim -7000$ au) are at least one order of magnitude larger than ours in the most similar geometry they considered (Li displaced about 1.5 Å from the center towards an hexagon). Campbells' values are based on orbitals obtained from a ROHF calculation, while ours are computed at the UB3LYP level. The large differences between the two results confirm the unreliability of the HF method for the hyperpolarizabilities of Li@C_{60}, as found here for the UHF values (cf. Table 4.1). In the same approximation, but now using RHF, they obtain $\sim$50.000 au for the diagonal component in C_{60} [6]. This value is about 2.5 times smaller than ours, which may be mostly due to the different basis sets, but also due to different approximations in both approaches, as well as differences in geometry etc. Finally, we note that Jansik et al. [21] computed values for the (hyper)polarizabilities of C_{60} in I_h symmetry with analytic response theory using larger basis sets, specifically tailored for the purpose of computing hyperpolarizabilities, and obtained $\alpha_{av} = 547.0$ au and $\gamma_{av} = 118 \times 10^3$ au with B3LYP using the cc-pVDZ+*spd* (their notation) basis set. These values are quite comparable to ours, taking into account the differences in symmetry (I_h vs. C_s), geometry, and basis set (it was assumed in our case that the geometry of C_{60} is sufficiently spherical so that the nondiagonal terms of γ do not deviate appreciably from the relation $\gamma_{iijj} = 1/3\gamma_{iiii}$).

We also investigated the influence of the position of the Li atom along the dipole axis on the electric properties of Li@C_{60}. The UB3LYP/6-31+G* values for different distances $r_{\text{Li-O}}$, where O denotes the center of mass of the cage, are shown in Table 4.3. For reference purposes the surface of the cage is at a distance of $\sim$3.4 Å. Interestingly, the diagonal z-component of the polarizability, first hyperpolarizability, and dipole moment change little (μ, α) or moderately (β) (for μ we are assuming that the value of the change is reliable even though the value of the dipole moment itself is not) between $r = 0.729$ and 2.0 Å, but γ_{zzzz} is altered much more drastically, even undergoing a sign change at small distances from the cage center. This is consistent with γ being due to electron density that is distant from the surface of the cage.

Table 4.3 Electronic dipole moment and (hyper)polarizabilities for Li@C_{60} along the dipole (z) axis as a function of the distance between the center of the cage and the Li atom ($r_{Li\text{-}O}$/au), computed at the UB3LYP/6-31+G* level

r_{Li-O}	μ_z^e	α_{zz}^e	β_{zzz}^e	$\gamma_{zzzz}^e(\times 10^3)$
0	−0.117	597.6	416	−86
0.729	0.159	594.4	1112	−19
0.958	0.243	592.8	1278	8
1.058	0.278	592.1	1340	20
1.158	0.310	591.4	1396	32
1.358	0.367	589.9	1492	55
1.458	0.390	589.0	1533	66
1.558	0.409	588.2	1569	76
1.658	0.424	587.3	1601	87
2.0	0.440	584.5	1676	123

The large gradient in γ_{zzzz} for Li@C_{60} could possibly be used in a potential nonlinear "flip-flop" device. This would require a mechanism such as an STM electric field to shift the equilibrium position of the Li atom between different regions. The magnitude of such a shift has been investigated by Delaney and Greer [2] who found that it is difficult to move the Li atom very far because of the large screening effect of the fullerene cage. In the calculations reported below we find that a shift from the field-free position of about $\Delta z \sim 0.03$ au will result when a 0.0128 au field is applied. According to Table 4.3, this shift is much too small to change γ_{zzzz} appreciably.

4.2.3 NR Contribution to Vibrational NLO Properties

In this section we present the nuclear relaxation (NR) contributions to the vibrational (hyper)polarizabilities of Li@C_{60} and [Li@C_{60}]$^+$. As previously stated our treatment requires a geometry optimization in the presence of a finite field. A problem can arise when there are multiple minima on the PES separated by low energy barriers. The finite field method works satisfactorily in that event as long as the field-dependent optimized structure corresponds to the same minimum as the field-free optimized structure. This was the case in previous work on ammonia [42], which has a double minimum potential. However, it is sometimes not the case for the endohedral fullerenes considered here, especially Li@C_{60}. In fact, we were unable to determine the NR contribution in the x direction, i.e. perpendicular to the symmetry plane, for that molecule. It was possible to obtain α_{xx}^{nr}, based on the alternative analytical formulation [32–34], utilizing field-free dipole (first) derivatives and the Hessian. The analytical polarizability components in the other two directions were, then, used to confirm the values of the corresponding finite field method for those properties.

Table 4.4 Nuclear relaxation (NR) contribution to the diagonal components of the static α, β and γ of $Li@C_{60}$ and $[Li@C_{60}]^+$ calculated at the (U)B3LYP/6-31G level and the UB3LYP/6-31+G level (in *square* brackets)

$i =$	$Li@C_{60}$			$[Li@C_{60}]^+$		
	x	y	z	x	y	z
α_{ii}^{nr}	14.7[a]	10.3	10.2 [11.9]	10.4	9.4	4.5
β_{iii}^{nr}	–[b]	−125.9	794.6 [(912–915)]	0	95	18
γ_{iiii}^{nr} x10^3	–[b]	−90	(25–40) [(52–81)]	560	190	37

[a]Computed analytically; see text
[b]Not determined; see text

In addition to the situation just discussed, it was also found that the electric field can sometimes lead to a change of electronic state, as detected by a sudden jump in the computed polarizability. This further limited the range of applicable field strengths and, thus, the range of properties that could be computed with sufficient statistical confidence.

Our results for the static NR contributions to the diagonal (hyper)polarizability components are shown in Table 4.4. Most of the values were obtained at the (U)B3LYP/6-31G level. For comparison, a few calculations were also done for $Li@C_{60}$ at the UB3LYP/6-31+G level. As seen from the Table, diffuse basis functions have a non-negligible effect on the computed values, although the effect is smaller than on the electronic properties (cf. Table 4.1). Note in particular that there is no *qualitative* change of any vibrational property upon going from 6-31G to 6-31+G, in contrast to the electronic properties where 6-31G gives a negative value for γ_{zzzz}, while it is positive for 6-31+G. Thus, we expect that the values of vibrational properties obtained with the 6-31G basis are qualitatively correct, although the accuracy becomes worse for γ^{nr} than it is for α^{nr} or β^{nr}.

For α the vibrational contributions are quite small in comparison with their electronic counterparts (cf. Table 4.2) . In the case of $[Li@C_{60}]^+$ this appears, at first glance, to contrast with what Whitehouse and Buckingham (WB) [4] have previously found. However, their values were obtained by classical averaging in the high temperature limit—in this case above 20 K—while ours are for 0 K. Another difference is that WB obtain the complete vibrational polarizability and second hyperpolarizability, albeit very approximately, whereas we have not included the so-called curvature contribution [36]. For the polarizability one would "normally" expect the latter to be substantially smaller than the NR term, but endohedral fullerenes are not "normal" molecules and that may not be the case here. Of the several approximations in the WB treatment, one of the most questionable is the spherical approximation for the field-free potential, which Zhang et al. have shown does not qualitatively reproduce the low energy vibrational spectrum of the neutral. This is not to mention that WB considered a rigid cage and motion along the C_{5v} symmetry axis, rather than the lower energy C_{3v} symmetry axis.

In contrast to α, the NR contributions to β and γ, are quite large. For the cation, in particular, two diagonal components of γ^{nr} are larger than the correspond-

Table 4.5 Components of the NR contribution to the dc-Pockels effect ($\beta^{nr}_{ijk}(-\omega;\omega,0)_{\omega\to\infty}$) computed for Li@$C_{60}$ in the infinite optical frequency approximation from Eq. 4.8, at the (U)B3LYP/6-31G level

$ij =$	xx	yy	yz	zz
β^{nr}_{ijy}	−23	−33	51	−9
β^{nr}_{ijz}	−199	119	15	200

ing electronic components. Such an increase in the relative magnitude of the NR (hyper)polarizabilities, as compared to the electronic values, as the order of non-linearity increases is often observed in conjugated systems [31, 43]. However, it is usually not to as large an extent as found here. The relationship with degree of non-linearity may be connected with the fact that only dipole derivatives enter into the linear NR polarizabilities, whereas β^{nr} and γ^{nr} depend additionally on polarizability derivatives, as the perturbation expressions for these quantities show [34]. Because of the conjugation the polarizability derivatives tend to be large (small changes in bond length alternation cause large changes in α). Furthermore, the higher-order vibrational polarizabilities (as opposed to the linear vibrational polarizability) depend upon electrical and mechanical anharmonicity which, undoubtedly, is especially large for the systems we are considering. WB give an expression for γ, but no numerical values. In addition, they considered only linear terms in their field-dependent vibrational Hamiltonian. This omits contributions to the vibrational hyperpolarizability that are generally important as suggested above.

In Table 4.5 we show the computed NR contribution to several components of the dc-Pockels effect for Li@C_{60}, i.e. $\beta^{nr}_{ijy}(-\omega;\omega,0)$ and $\beta^{nr}_{ijz}(-\omega;\omega,0)$ where $ij = xx$, yy, yz, and zz. (see Eq. 4.8). Comparison with the corresponding electronic values of Table 4.2 shows that the vibrational contribution is relatively small, compared to the corresponding static electronic component, but not negligible. The values obtained here may be compared with those of a typical donor–acceptor molecule, H_2N–$(CH{=}CH)_3$–NO_2, for which the ratio $\beta^{nr}_{iii}(-\omega;\omega,0)_{\omega\to\infty}/\beta^{e}_{iii}(0;0,0)$ along the dipole direction was found to be about 0.7 at the RHF level and about 0.2 at the MP2 level [43]. For LiC_{60} this ratio is 0.13 at the UB3LYP level.

4.3 Conclusions

We have computed both electronic and NR vibrational contributions to the (hyper) polarizabilities of the prototype endohedral fullerene Li@C_{60} and its cation. A number of these properties were obtained for the first time. In other cases our results differ quite signicantly from those previously determined using more approximate approaches. The latter include the static electronic properties calculated by Campbell et al. [5, 6]. Although, for the cation, there is a large difference between our values of the static vibrational contribution to α and those reported by Whitehouse and

Buckingham [4], these results are not really comparable because their calculations include the effect of temperature. In addition, they applied several strong approximations, such as assuming a spherical field-free potential inside the cage. On the other hand, our calculations do not include higher-order vibrational contributions omitted in the NR treatment. It would be worthwhile to add temperature-dependence to the NR approach as we plan to do in the future. Whereas the NR contribution to the static α is quite small for both endohedral fullerenes, it becomes quite large for the static hyperpolarizabilities. This contribution is reduced for the dynamic Pockels effect, computed in the infinite optical frequency approximation, but is still not negligible.

For $[Li@C_{60}]^+$ the calculated (hyper)polarizabilities are roughly comparable to those of the hypothetical non-interacting system obtained by charge transfer of the Li valence electron to the cage giving $Li^+ + C_{60}$. The same is true for the linear polarizability of the neutral but the non-interacting charge transfer model completely breaks down for the hyperpolarizabilities.

We consider our work as a substantial step towards the final goal of a full computational characterization of the linear and nonlinear electric dipole properties of endohedral fullerenes. As far as vibrational contributions are concerned, in addition to the NR treatment of low-order perturbation terms, there is an established method for obtaining all remaining contributions through calculation of zero-point vibrationally averaged properties at the relaxed field-dependent geometry [36]. What is needed is a robust procedure for carrying out the geometry optimizations when the PES has multiple minima and/or other strongly anharmonic features. Work is in progress on a reduced dimensionality scheme that may be combined with quasidegenerate perturbation theory to treat such circumstances [42].

References

1. Shinohara, H.: Rep. Prog. Phys. **63**, 843 (2000)
2. Delany, P., Greer, J.C.: Appl. Phys. Lett. **84**, 431 (2004)
3. Ross, R.B., Cardona, C.M., Guldi, D.M., Sankaranarayanan, S.G., Reese, M.O.: Kopidakis, N., Peet, J., Walker, B., Bazan, G.C., Van Keuren, E., Holloway, B.C., Drees, M. Nat. Mater. **8**, 208 (2009)
4. Whitehouse, D.B., Buckingham, A.D.: Chem. Phys. Lett. **207**, 332 (1993)
5. Campbell, E.E.B., Fanti, M., Hertel, I.V., Mitzner, R.: Zerbetto. F. Chem. Phys. Lett. **28**(8), 131 (1998)
6. Campbell, E.E.B., Couris, S., Fanti, M., Koudoumas, E., Krawez, N., Zerbetto, F.: Adv. Mater. **11**, 405 (1999)
7. Yaghobi, M., Rafie, R., Koohi, A.: J. Mol. Struct. Theochem. **905**, 48 (2009)
8. Yaghobi, M., Koohi, A.: Mol. Phys. **108**, 119 (2010)
9. Liu, J., Iwata, S., Gu, B.: J. Phys. Condes. Matter. **6**, L253 (1994)
10. Tang, C.M., Zhu, W.H., Deng, K.M.: J. Molec. Struct. Theochem. **894**, 112 (2009)
11. Tang, C.M., Fu, S.Y., Deng, K.M., Yuan, Y.B., Tan, W.S., Huang, D.C., Wang, X.: J. Mol. Struct. (Theochem) **867**, 111 (2008)
12. Torrens, F.: J. Phys. Org. Chem. **15**, 742 (2002)
13. Torrens, F.: Nanotechnology **13**, 433 (2002)
14. Yan, H., Yu, S., Wang, X., He, Y., Huang, W., Yang, M.: Chem. Phys. Lett. **456**, 223 (2008)

15. He, J., Wu, K., Sa, R., Li, Q., Wei, Y.: Chem. Phys. Lett **475**, 73 (2009)
16. Pederson, M.R., Baruah, T., Allen, P.B., Schmidt, C.: J. Chem. Theory Comput. **1**, 590 (2005)
17. Kurtz, H.A., Stewart, J.J.P., Dieter, K.M.: J. Comput. Chem. **11**, 82 (1990)
18. Davia, P.J., Rabinowitz, P.: Numerical Integration, p. 166. Blaisdell, London (1967)
19. Gaussian 03, Frisch, M.J., Trucks, G.W., Schlegel, H.B., Scuseria, G.E., Robb, M.A., Cheeseman, J.R., Montgomery, J.A., Vreven, T.Jr., Kudin, K.N., Burant, J.C., Millam, J.M., Iyengar, S.S., Tomasi, J., Barone, V., Mennucci, B., Cossi, M., Scalmani, G., Rega, N., Petersson, G.A., Nakatsuji, H., Hada, M., Ehara, M., Toyota, K., Fukuda, R., Hasegawa, J., Ishida, M., Nakajima, T., Honda, Y., Kitao, O., Nakai, H., Klene, M., Li, X., Knox, J.E., Hratchian, H.P., Cross, J.B., Adamo, C., Jaramillo, J., Gomperts, R., Stratmann, R.E., Yazyev, O., Austin, A.J., Cammi, R., Pomelli, C., Ochterski, J.W., Ayala, P.Y., Morokuma, K., Voth, G.A., Salvador, P., Dannenberg, J.J., Zakrzewski, V.G., Dapprich, S., Daniels, A.D., Strain, M.C., Farkas, O., Malick, D.K., Rabuck, A.D., Raghavachari, K., Foresman, J.B., Ortiz, J.V., Cui, Q., Baboul, A.G., Clifford, S., Cioslowski, J., Stefanov, B.B., Liu, G., Liashenko, A., Piskorz, P., Komaromi, I., Martin, R.L., Fox, D.J., Keith, T., Al-Laham, M.A., Peng, C.Y., Nanayakkara, A., Challacombe, M., Gill, P. M.W., Johnson, B., Chen, W., Wong, M.W., Gonzalez, C., and Pople, J.A.: Gaussian, Inc., Pittsburgh, PA (2003)
20. Gaussian 09, Revision A.02, Frisch, M.J., Trucks, G.W., Schlegel, H.B., Scuseria, G.E., Robb, M.A., Cheeseman, J.R., Scalmani G., Barone, V., Mennucci, B., Petersson, G.A., Nakatsuji, H., Caricato, M., Li, X., Hratchian, H. P., Izmaylov, A.F., Bloino, J., Zheng, G., Sonnenberg, J. L., Hada, M., Ehara, M., Toyota, K., Fukuda, R., Hasegawa, J., Ishida, M., Nakajima, T., Honda, Y., Kitao, O., Nakai, H., Vreven, T., Montgomery, J.A., Peralta, J.E. Jr., Ogliaro, F., Bearpark, M., Heyd, J.J., Brothers, E., Kudin, K.N., Staroverov, V.N., Kobayashi, R., Normand, J., Raghavachari, K., Rendell, A., Burant, J.C., Iyengar, S.S., Tomasi, J., Cossi, M., Rega, N., Millam, J.M., Klene, M., Knox, J.E., Cross, J.B., Bakken, V., Adamo, C., Jaramillo, J., Gomperts, R., Stratmann, R. E., Yazyev, O., Austin, A.J., Cammi, R., Pomelli, C., Ochterski, J.W., Martin, R.L., Morokuma, K., Zakrzewski, V.G., Voth, G.A., Salvador, P., Dannenberg, J.J., Dapprich, S., Daniels, A.D., Farkas, O., Foresman, J.B., Ortiz, J.V., Cioslowski, J., and Fox, D.J.: Gaussian, Inc., Wallingford, CT (2009)
21. Jansik, B., Sałek, P., Jonsson, D., Vahtras, O., Ågren, H.: J. Chem. Phys. **122**, 054107 (2005)
22. Dalton, a molecular electronic structure program. Release 2.0, see http://www.kjemi.uio.no/software/dalton/dalton.html (2005)
23. Rinkevicius, Z., Jha, P.C., Oprea, C.I., Vahtras, O., Ågren, H.: J. Chem. Phys. **127**, 114101 (2007)
24. Jha, P.C., Rinkevicius, Z., Ågren, H.: ChemPhysChem. **10**, 817 (2009)
25. Christiansen, O.: Phys. Chem. Chem. Phys. **9**, 2942 (2007)
26. Hansen, M.B., Christiansen, O., Hättig, C.: J. Chem. Phys. **131**, 154101 (2009)
27. Bishop, D.M., Hasan, M., Kirtman, B.: J. Chem. Phys. **103**, 4157 (1995)
28. Luis, J.M., Duran, M., Andrés, J.L., Champagne, B., Kirtman, B.J.: Chem. Phys. **111**, 875 (1999)
29. Bishop, D.M., Dalskov, E.K.: J. Chem. Phys. **104**, 1004 (1996)
30. Quinet, O., Champagne, B.: J. Chem. Phys. **109**, 10594 (1998)
31. Luis, J.M., Duran, M., Kirtman, B.: J. Chem. Phys. **115**, 4473 (2001)
32. Bishop, D., Kirtman, B.: J. Chem. Phys. **95**, 2646 (1991)
33. Bishop, D., Kirtman, B.: J. Chem. Phys. **97**, 5255 (1992)
34. Bishop, D., Luis, J.M., Kirtman, B.: J. Chem. Phys. **108**, 10013 (1998)
35. Luis, J.M., Martí, J., Duran, M., Andrés, J.L., Kirtman, B.: J. Chem. Phys. **108**, 4123 (1998)
36. Kirtman, B., Luis, J.M., Bishop, D.M.: J. Chem. Phys. **108**, 10008 (1998)
37. Zhang, M., Harding, L.B., Gray, S.K., Rice, S.A.: J. Phys. Chem. A **112**, 5478 (2008)
38. Hernández-Rojas, J., Bretón, J.: Gomez Llorente, J. M. Chem. Phys. Lett. **235**, 160 (1995)
39. Hernández-Rojas, J.: Bretón, Gomez Llorente, J. M. Chem. Phys. Lett. **243**, 587 (1995)
40. Li, Y.S., Tománek, D.: Chem. Phys. Lett. **221**, 453 (1994)
41. Fowler, P.W., Madden, P.A.: Phys. Rev B **30**, 6131 (1984)
42. Luis, J.M., Reis, H., Papadopoulos, M.G., Kirtman, B.: J. Chem. Phys. **131**, 034116 (2009)
43. Luis, J.M., Champagne, B., Kirtman, B.: Int. J. Quant. Chem. **80**, 471 (2000)

Chapter 5
Fullerene–Porphyrin Dyads

5.1 Polyalkyne Chained Dyads

A design of organic molecules for a wide range of nonlinear optical applications is still the subject of intense experimental and theoretical research [1–3]. The ease of processing and modification as well as relatively low cost of production make organic materials promising candidates for future photonic applications. Depending on the destination, materials composed of organic molecules (Langmuir-Blodgett films, functionalized polymers, etc.) should fulfill certain criteria. Large value of nonlinear susceptibility ($\chi^{(2)}$, $\chi^{(3)}$, ...) is not the most important factor determining the applicability of a given material. The spectral characteristics is of similar significance in most applications. Different are spectral requirements for the two-photon absorption and for the frequency tripling. Computational techniques of electronic structure theory, in principle, can predict both linear as well as nonlinear optical properties of molecular systems. The accuracy of calculated properties depend much on the applied level of theory which, in turn, is closely related to the size of the investigated system. Density functional theory is nowadays commonly used as an alternative to computationally much more expensive electron correlation approaches. In particular, time-dependent density functional theory (TD-DFT) is the most frequent choice for studying of electronic excited states for molecules composed of more than 20 atoms [4]. It is well recognized, however, that most of the conventional functionals poorly reproduce the spectra of the charge-transfer complexes as well as have difficulties in correct prediction of excitation energies to Rydberg states [4]. Another important drawback of traditional functionals is the inability to accurately determine molecular (hyper)polarizabilities of long-chain molecules. Recently proposed long-range corrected (LRC) functionals, like LC-BLYP [5] or CAM-B3LYP [6], are expected to cure above mentioned deficiencies of DFT. The LRC functionals, however, need to be tested for large variety of molecular systems before they may be applied on regular basis for prediction of various properties. As a part of the present study we shall apply recently proposed LC-BLYP functional, together with other

O. Loboda, *Quantum-Chemical Studies on Porphyrins, Fullerenes and Carbon Nanostructures*, Carbon Nanostructures, DOI: 10.1007/978-3-642-31845-0_5,

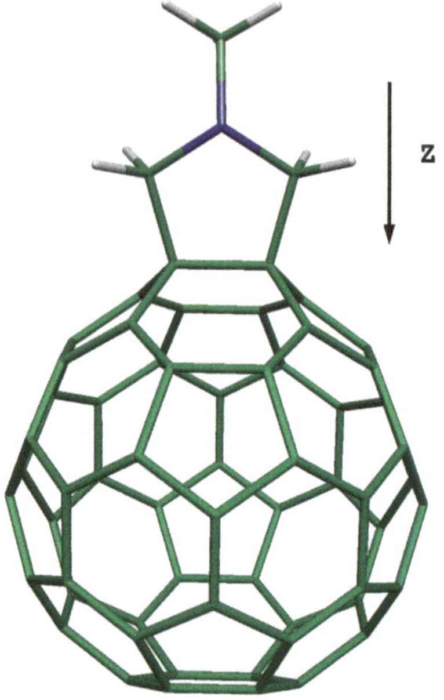

Fig. 5.1 Structure of *N*-methylfulleropyrrolidine

functionals, to compute linear and nonlinear optical properties of model [60]fullerene derivatives. The critical assessment, as a part of the calibration procedure for this class of compounds, will be performed by comparing the results of calculations either with existing experimental data or the values computed using MP2 method. The interest in [60]fullerene derivatives is due to their photophysical properties which make them promising candidates in photonic applications. In this preliminary study, we focus on various [60]fullerene–porphyrin dyads of donor–acceptor (D–A) character in order to analyze the structure-property relations for this class of compounds [7, 8, 1] (Fig. 5.1).

5.1.1 Computational Details

The averaged (hyper)polarizabilities we use throughout this study are defined as in Ref. [10]. The calculations of the linear spectra were performed using the time-dependent density functional theory (TD-DFT) with the aid of the Gaussian 03 package [11]. We also used linear-scaling approach as implemented in the SIESTA program [12]. In order to calculate the photoabsorption cross section $\sigma(\omega)$, the dipole strength function were evaluated according to the expression:

$$S(\omega) = \frac{2\omega}{\pi}\text{Im}\frac{\text{Tr}\alpha_{ij}(\omega)}{3} \tag{5.1}$$

where Imα stands for the imaginary part of polarizability which, in turn, is given by:

$$\text{Im}\alpha(\omega) = \omega \frac{\text{Re}\, D(\omega)}{E}. \tag{5.2}$$

$D(\omega)$ is the Fourier transform of the dipole moment $D(t)$ which is calculated at every step from the electron density [13]:

$$D(\omega) = \int e^{i\omega t - \delta t} D(t) \mathrm{d}t. \tag{5.3}$$

The geometry optimizations of molecules discussed here were performed using the semiempirical PM3 Hamiltonian [14]. The linear and nonlinear optical properties were calculated either by numerical differentiation of total energy given by Eq. (1) or, in the case of PM3 and PM6 methods [14, 15], analytically. Semiempirical computations were performed with the aid of the MOPAC program [16]. At the DFT, HF and MP2 levels of theory we used fast multipole method to speed up the calculations [17]. In particular, the LC-BLYP calculations were performed using the GAMESS US program [18]. All the properties are expressed in atomic units.

5.1.2 Results and Discussion

In the present section we shall analyze both the excitation spectra of *N*-methylfullero-pyrrolidine (see Fig. 5.2) as well as its nonlinear optical properties. *N*-methylfullero-pyrrolidine is chosen as a model system of donor–acceptor type. Figure 5.2 presents the spectra of the system measured in cyclohexane [19] which has very small value of dielectric constant. Thus, we compare the experimental data with the calculated spectra for the system *in vacuo*. The results of computations are also presented in Fig. 5.2. The experimental maxima are located near 300, 250 and 210 nm. We see that the spectra simulated at the PBE0/DZP level of theory using SIESTA program is very similar to the experimental one. The location of the maxima near 210 nm is accurately determined. The other experimentally observed maxima near 250 nm is predicted to be at 228 nm. The location of this band calculated using B3LYP functional is found to be at 233 nm. Based on the data presented in Fig. 5.2 we conclude that the overall agreement between theoretical and experimental spectra is satisfactory. The small discrepancies can also be attributed partially to the neglect of solvent effects in our simulations. It is well known that the higher the polarity of the excited state, the larger is the excitation energy in polar solvent as compared to the gas phase. The values of dipole moment differences along *z*-axis for states of largest intensity are presented in Fig 5.2. The calculations were performed at the PBE0/6-31G(d) level of theory by numerical differentiation of excitation energies with respect to the electric field. As we see, the dipole moment difference between the excited and the ground state can be as large as 3.5 D.

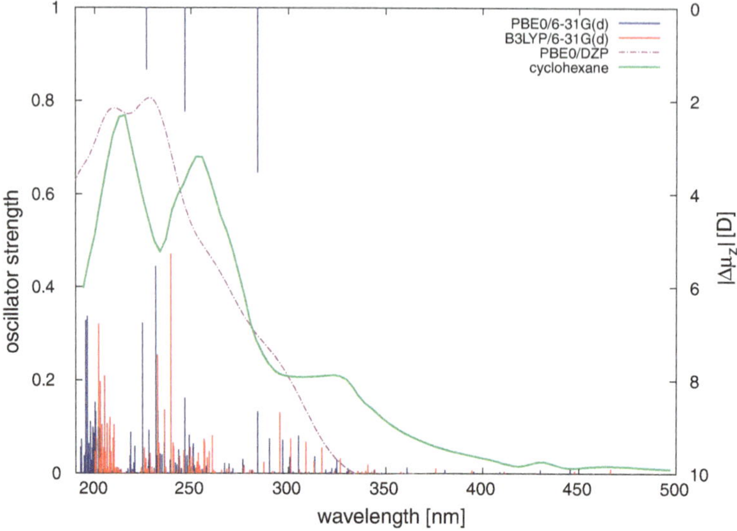

Fig. 5.2 One-photon absorption spectra of *N*-methylfulleropyrrolidine

Table 5.1 Linear and nonlinear optical properties of *N*-methylfulleropyrrolidine

	μ_z	α_{zz}	β_{zzz}	γ_{zzzz}
HF/6-31G(d)	−1.14	542.8	−7	$47{\cdot}10^3$
MP2/6-31G(d)	−1.23	573.3	−146	$97{\cdot}10^3$
BLYP/6-31G(d)	−1.14	610.2	−3489	$1176{\cdot}10^3$
LC-BLYP/6-31G(d)	−1.08	554.9	−180	$84{\cdot}10^3$
LC-BLYP/Z3Pol	−1.24	636.1	–	–
PM3	−1.12	503.2	−286	$106{\cdot}10^3$
PM6	−1.49	616.9	−363	$105{\cdot}10^3$

All values are given in atomic units

The results of calculations of dipole moment as well as molecular (hyper)polarizabilities for *N*-methylfulleropyrrolidine are presented in Table 5.1. The comparison of the values computed at the HF and MP2 levels of theory allows for the direct assessment of the influence of electron correlation on α, β and γ. We clearly see that both β_{zzz} as well as γ_{zzzz} substantially increase upon the inclusion of electron correlation effects. The absolute values of electrical properties are perhaps underestimated, as the employed basis set (6-31-G(d)) lacks diffuse functions, which are known to be necessary for reliable prediction of molecular (hyper)polarizabilities. This is demonstrated by the value of polarizability calculated using Z3Pol basis set, which is the recently proposed basis set for accurate prediction of α [20]. Indeed, we see that the Z3Pol basis predict the value of α to be 15 % larger than the value calculated using 6-31G(d) basis set. Another important fact that clearly emerges from Table 5.1 is the inability of the BLYP functional to correctly predict values of β_{zzz} and γ_{zzzz}

Table 5.2 The dipole moment (μ), average polarizability ($\overline{\alpha}$) and first hyperpolarizability ($\overline{\beta}$) for two model [60]fullerene derivatives

Molecule	μ	$\overline{\alpha}$	$\overline{\beta}$	Method
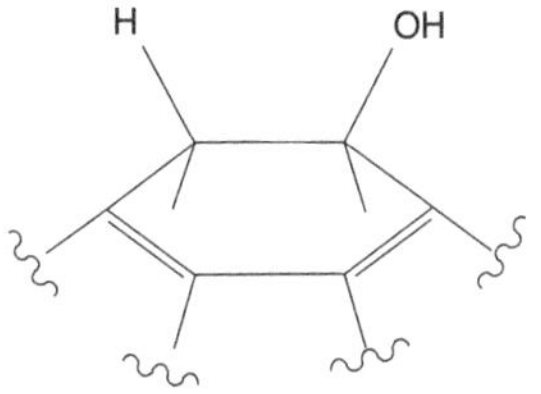 H, OH	0.71	452.94	−26.8	HF/6-31G(d)
	0.66	485.30	−48.4	BLYP/6-31G(d)
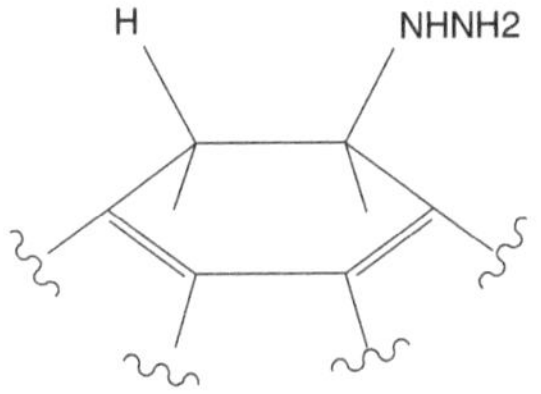 H, NHNH2	0.95	466.27	−23.1	HF/6-31G(d)
	1.03	510.28	5559.5	BLYP/6-31G(d)

All values are given in atomic units

for the studied system. The BLYP functional predicts the values of hyperpolarizabilities to be over order of magnitude larger than these computed on other levels of theory. The overshoot problem for the considered system is quite striking as it is not observed for γ of [60]fullerene. The LC-BLYP functional successfully cure the overshoot problem. Both β_{zzz} and γ_{zzzz} are very close to the MP2 value which is considered here to be the reference point. Finally, we note that molecular hyperpolarizabilities are predicted reasonably well by PM3 and PM6 methods. Thus, we shall use these two methods for further computations in the present study. In order to gain more insight into the overshoot problem we performed additional calculations for two model systems presented in Table 5.2. The [60]fullerene derivatives were selected according to the value of Hammett σ_p constant which may be a semiquantitative measure of electron donating capabilities of substitutes [21]. It is not the purpose to perform the correlation analysis between the overshoot and the Hammett σ_p constant. Instead, we rather use σ_p as a parameter that allows us for ordering the systems, i.e. OH group ($\sigma_p = -0.37$) is considered to be much weaker donor than the $NHNH_2$ group ($\sigma_p = -0.55$). As it is seen, in the case of strong electron donating substituent the overshoot for first hyperpolarizability is even larger than that for *N*-methylfulleropyrrolidine.

As it was reported earlier in the subject literature, the NLO values of [60]fullerene-dyads calculated with the aid of the PM3 method show satisfactory agreement with the DFT results and experimental third harmonic generation values [1]. Here, we also use the PM6 method [15]. The studied systems are presented in Fig. 5.3. The top structure is [60]fullerene-dyad composed of fullerene and free base porphyrin linked by conjugated linear carbon spacer. In the middle structure, the free base por-

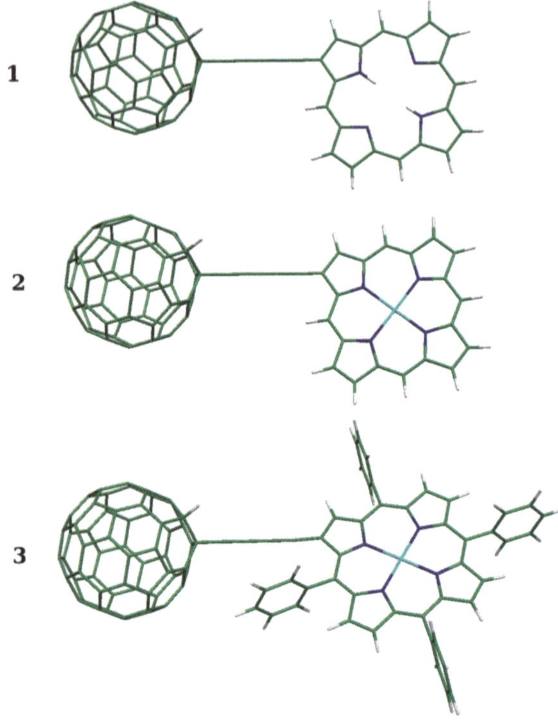

Fig. 5.3 The structure of investigated [60]fullerene–porphyrin dyads

phyrin (H_2P) is substituted by zinc porphyrin (ZnP). Finally, the bottom structure (ZnTPP-C_4-[60]) represents modified ZnP in which four bridging methine protons are substituted by phenyl rings. The results of PM3 and PM6 calculations of L&NLO properties are given in Table 5.3. From the Table 5.3 it is clear that both PM3 and PM6 methods predict substantial increase of L&NLO properties upon complexation of porphyrin with zinc atom and phenyl ring substitution. This effect can be attributed to the enhancement of the donor–acceptor character of [60]fullerene–porphyrin dyad. Another interesting point is the effect of the size of the conjugation bridge on L&NLO properties of fullerene–porphyrin dyads. Table 5.4 contains the results of calculations of dipole moment, polarizability and second hyperpolarizability of H_2P-$(C)_n$-[60] dyad for the different length of the conjugated chain, which links the fullerene and the zinc porphyrin units. One can see from the Table 5.4 that the polarizability and second hyperpolarizability increase almost linearly with the π-conjugated carbon chain length, i.e. the Pearson's product-moment correlation coefficient for γ, calculated using the PM6 model, is equal to 0.988 (Fig. 5.4).

Table 5.3 The averaged values of linear and nonlinear optical properties of H_2P-C_4-[60], ZnP-C_4-[60], ZnTPP-C_4-[60]

	Method	H_2P-C_4-[60]	ZnP-C_4-[60]	ZnTPP-C_4-[60]
$\overline{\mu}$	PM3	1.95	2.10	2.12
	PM6	3.17	3.11	2.92
$\overline{\alpha}$	PM3	835.03	869.88	1130.32
	PM6	1003.59	1028.88	1352.82
$\overline{\beta}$	PM3	695.22	2310.55	3718.38
	PM6	676.23	2234.73	5931.43
$\overline{\gamma}$	PM3	724846.9	714525.7	917296.6
	PM6	625658.8	763584.6	922977.8

All values are given in atomic units

Table 5.4 The comparison of polarizability (α) and second hyperpolarizability (γ) of H_2P-$(C)_n$-[60] dyad, where n= 0,2,4,6,8

	n	0	2	4	6	8
$\overline{\mu}$	PM3	1.19	1.52	1.95	1.47	1.55
	PM6	1.81	2.12	3.17	2.13	2.32
$\overline{\alpha}$	PM3	733.86	773.47	835.03	871.70	924.81
	PM6	875.92	926.29	1003.59	1054.46	1127.62
$\overline{\gamma}$	PM3	438535.3	530326.8	724846.9	827147.1	1030056.6
	PM6	435212.9	585509.6	625658.8	804858.5	901431.6

All values are given in atomic units

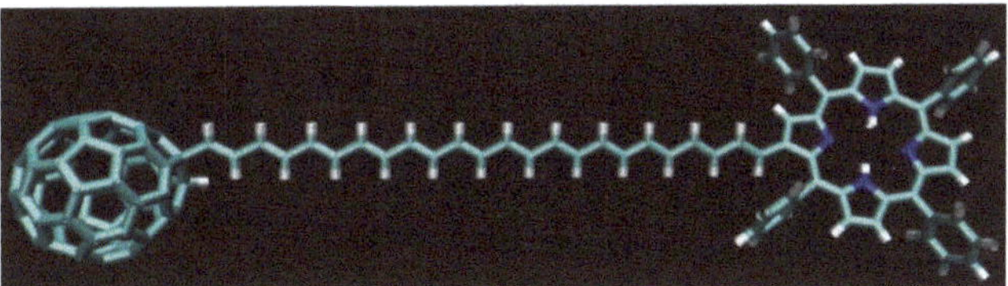

Fig. 5.4 The structural motif of investigated molecules

5.1.3 Conclusions

The structure(Zn,Ph)-property(α,β,γ,) relationship has been investigated in this work by semiempirical PM3 and PM6 methods. The ZnTPP-$(C)_4$-[60] dyad was found to exhibit large values of second hyperpolarizability. We found almost linear correlation between the size of the conjugated polyalkynyl linkage and the optical properties of the whole fullerene–porphyrin dyad. As a part of the present study, we have found that the nonlinear optical properties of model donor-substituted [60]fullerene derivatives are poorly described by the BLYP functional. The long-range corrected LC-BLYP functional, on the other hand, successfully cures the overshoot problem both for β and γ. The calculations for *N*-methylfulleropyrrolidine shows that the linear absorption spectra is reproduced quite well by conventional PBE0 and B3LYP functionals.

5.2 Polyacetylene Chained Dyads

The increasing demand for faster data processing, storage and distribution can only be fulfilled by ongoing minimization of the basic electronic devices. The traditional silicon-based technologies used nowadays are approaching intrinsic limits in this respect, and new approaches are needed. Photonic technology, where light is used as information carrier instead of electrons, is considered to offer the answer. An important step towards this goal is the development of new photonic materials with large NLO properties. Fullerene-based nano-hybrids as well as carbon heterostructured nanotubes are considered to be a highly promising class of such materials. Linear scaling method, O(N), are among the most reliable and cost efficient approaches available for the calculation of the properties of large systems. Fullerene derivatives with appropriate donors have significant second-order nonlinearity. It is known that fullerenes are excellent acceptors.

Conventional ab initio methods require high computational resources, even the Hartree-Fock approximation formally scales as M^4, where M is the number of basis functions. Thus various linear scaling methods and algorithms have been developed and applied in a large variety of fields. Unfortunately, very little is known on the performance of such approaches on the computation of the NLO properties. The main target of the present work is design of nano-materials for photonic applications. The key parameters for such a design are the nonlinear optical (NLO) properties. These properties have two contributions, the electronic and the vibrational. The vibrational contributions are usually negligible and therefore out of the scope of the present book. For the evaluation of electronic contribution we used finite field methodology.

5.2.1 Results

An essential aspect of this book is to check the performance of the linear scaling elongation method for the computation of the NLO properties of the nanotubes and fullerene-based materials.

We improved the performance of elongation method by refining the localization scheme. After the regional localization procedure is done there we have two sets of occupied and vacant orbitals assigned to A and B regions. However under closer consideration one can notice the admixture of B region orbitals in frozen A region and vise versa. This may lead to low accuracy in the calculation of the total energy for the whole system. To circumvent this problem we move such kind of orbitals with significant $\langle A|B\rangle$ overlap from A region into B region, where in the next elongation step they will be localized again and placed into A region at the most. In this way we maintain the constant number of "transferred" orbitals and avoid tailings of orbitals from the active region in the A part.

Table 5.5 The averaged values of linear and nonlinear optical properties of $H_2TPP\text{-}C_4H_4\text{-}[60]$

Molecule	Method	μ_z	α_{zz}	β_{zzz}
$H_2TPP\text{-}C_4H_4\text{-}[60]$	PM_6	−4.110	1738.28	−10232.8
$H_2TPP\text{-}C_4H_4\text{-}[60]$	HF[a]	−5.413	1556.87	−4096.5
$H_2TPP\text{-}C_4H_4\text{-}[60]$	HF	−5.413	1556.88	−4097.0
$H_2TPP\text{-}(C_4H_4)_2\text{-}[60]$	HF	−5.763	1836.85	−6672.1
$H_2TPP\text{-}(C_4H_4)_3\text{-}[60]$	HF	−5.941	2152.85	−9925.7
$H_2TPP\text{-}(C_4H_4)_4\text{-}[60]$	HF	−6.033	2479.35	−13445.0
$H_2TPP\text{-}(C_4H_4)_5\text{-}[60]$	HF	−6.085	2917.02	−14847.1
$H_2TPP\text{-}(C_4H_4)_6\text{-}[60]$	HF	−6.111	3142.20	−17667.4

μ-permanent dipole moment (debye), α-polarizability, β, first-order hyperpolarizability. Hyper/polarizability values are given in atomic units
[a]Conventional Calculations

This development allowed to increase the accuracy of elongation calculation in several orders of magnitude especially in the cases when the delocalized π-extended systems are treated.

As a starting point, we used the module, which involves the dyad of porphyrin and [60]fullerene linked with an alkene chain. These systems represent the first example of a new class of donor–acceptor derivatives in which π-conjugation extends from the porphyrin ring system directly to the fullerene surface. From the Table 5.5 it is clear that nature of molecular bridge which links fullerene and free-base porphin units plays substantial role in L&NLO properties of entire fullerene–porphin system. The polyalkene chain, enriched with π electrons provides better conductivity within fullerene-chromophore dyad. One can see from the Table 5.5 that the Elongation Method results are in an excellent agreement with the conventional calculations. The HF values are presented only for z component since the molecules have been oriented along z dipole moment. In comparison with the HF results PM6 method tends to overestimate first hyperpolarizability value polarizability however it has a qualitative agreement with "ab initio" polarizability value look Table 5.5.

5.2.2 Conclusions

The structure-property relationship has been investigated in this work by linear scaling Elongation HF method. The $H_2TPP\text{-}(C_4H_4)_n\text{-}[60]$ dyad was found to be the perspective candidate for further photonic applications. The post-localization rearrangement of RLMO's has been implemented in developmental version of elongation method and tested on both nanotube and fullerene–porphyrin dyads. The results of the present work show that the accuracy of elongation calculation improved in two orders of magnitude. Thus the elongation method is proved to be very useful for studying NLO properties of highly conjugated nano-systems.

References

1. Xenogiannopoulou, E., Medved, M., Iliopoulos, K., Couris, S., Papadopoulos, M.G., Bonifazi, D., Sooambar, C., Mateo-Alonso, A., Prato, M.: ChemPysChem. **8**, 1056 (2007)
2. Hamasaki, R., Ito, M., Lamrani, M., Mitsuishi, M., Miyashita, T., Yamamoto, Y.: J. Mater. Chem. **13**, 21 (2003)
3. Barbosa, A.G.H., Nascimento, M.A.C.: Chem. Phys. Lett. **343**, 15 (2001)
4. Dreuw, A., Head-Gordon, M.: Chem. Rev. **105**, 4009 (2005)
5. Iikura, H., Tsuneda, T., Yanai, T., Hirao, K.: J. Chem. Phys. **115**, 3540 (2001)
6. Yanai, T., Tew, D., Handy, N.C.: Chem. Phys. Lett. **393**, 51 (2004)
7. Galloni, P., Floris, B., De Cola, L., Cecchetto, E., Williams, R.M.: J. Phys. Chem. C **111**, 1517 (2007)
8. Vail, S.A., Schuster, D.I., Guldi, D.M., Isosomppi, M., Tkachenko, N., Lemmetyinen, H., Palkar, A., Echegoyen, L., Chen, X., Zhang, J.Z.H.: J. Phys. Chem. B **110**, 14155 (2006)
9. Sim, F., Chin, S., Dupuis, M., Rice, J.E.: J. Phys. Chem. **97**, 1158 (1993)
10. Bishop, D.M., Norman, P.: In: Nalwa, H.S. (ed.) Handbook of Advanced Electronic and Photonic Materials and Devices, Academic Press, San Diego (2001) (And references therein)
11. Frisch, M.J., et al.: Gaussian 03, Rev. D02, Gaussian, Inc., Wallingford (2004)
12. Sánchez-Portal, D., Ordejón, P., Artacho, E., Soler, J.M.: Int. J. Quant. Chem. **65**, 453 (1997)
13. Tsolakidis, A., Kaxiras, E.: J. Phys. Chem. A **109**, 2373 (2005)
14. Stewart, J.J.P.: J. Comput. Chem. **10**, 221 (1989)
15. Stewart, J.J.P.: J. Comput. Chem. **10**, 209 (1989)
16. MOPAC2007, J.J.P. Stewart Computational Chemistry, Version 7.221L web: http://OpenMOPAC.net
17. Millam, J.M., Scuseria, G.E.: J. Chem. Phys. **106**, 5569 (1997)
18. Schmidt, M.W., et al.: J. Comput. Chem. **14**, 1347 (1993)
19. Maggini, M., Scorrano, G.: J. Am. Chem. Soc. **115**, 9798 (1993)
20. Benkova, Z., Sadlej, A.J., Oakes, R.E., Bell, S.J.: J. Comput. Chem. **26**, 145 (2005)
21. Hansch, C., Leo, A., Taft, R.W.: Chem. Rev. **91**, 165 (1991)

Chapter 6
Linear Scaling Methodology

In 1991 the elongation method, an efficient method for quantum mechanical calculations of large systems, was originally proposed by Imamura [1] during one of his stays in Heidelberg, Germany. Although in the early 1990s the concept of and need for order-N [O(N)] methods did not exist, Prof. Akira Imamura was thinking about "how to avoid direct SCF calculations for large biological systems (biopolymers composed of hundreds if not thousands of residues of amino acids or nucleic acid base pairs in proteins and DNA or RNA) by treating only the local interactions between a few neighbor residues in large systems." While contemplating how to perform such calculations, he got the idea of "theoretically simulating the synthesis of polymers so as to mimic the chemical reactions which occur in nature during polymerization reactions to form peptides, proteins and polynucleotides. Hence he named the method he developed the elongation method, due to the nature of the first so-called buildup of the model from calculations for monomeric units. By doing so, he developed one of the first O(N) methods. First, he developed the method conceptually and then he coded it at the extended Hückel level [1] and was then able to confirm that it was working well and achieved very good accuracy in comparison with the more expensive conventional methods. Subsequently, in his group at Hiroshima University he continued to develop and extend this method at the ab initio single determinant Hartree–Fock level of theory [2], at the Kohn–Sham density functional theory (KS-DFT) level of theory [3], and subsequently, at Kyushu University, including electron correlation effects at the ab initio wave function theory (WFT) at the Moller–Plesset perturbation theory to second order (MP2) [4]. This is in line with the early work of Professor Imamura on the development of rules for chemical reactions for both ground and excited electronic state chemical reactions, the so-called Woodward–Hoffmann rules, which are necessary to understand in order to gain a better understanding of physical and chemical properties of large biological systems. For a good review of the various extensions to the original elongation method which have been developed up until 2006, see Ref. [5].

Recent progress in developing linear scaling electronic structure methods has lead to the development of a variety of O(N) techniques [6, 7]. The Fermi operator

O. Loboda, *Quantum-Chemical Studies on Porphyrins, Fullerenes and Carbon Nanostructures*, Carbon Nanostructures, DOI: 10.1007/978-3-642-31845-0_6,

expansion/projection (FOP/FOE) [8, 9], Density-Matrix Minimization methods [10] speak for themselves. The basic idea of Divide-and-Conquer [11] method is to calculate certain regions of the density matrix by considering sub-volumes and then to generate the full density matrix by adding these parts with the appropriate weights. The Orbital Minimization method [12] unlike Density-Matrix Minimization does not calculate the density matrix directly, but expresses it via Wannier functions. In the Optimal Basis Density-Matrix Minimization method [13] the fundamental basis functions are contracted in a first turn, then the Hamiltonian and overlap matrix are constructed in this new smaller basis. Many molecular properties remain for which to date no linear-scaling methods have been devised and implemented. The results for some molecular properties or electron correlation methods depend strongly on the size of the basis set. For example, post-Hartree–Fock methods require larger basis sets. This requirement rules out, in particular, the Density-Matrix Minimization method which becomes inefficient if the number of basis functions per atom is very large. In this respect it should be noted that even if a method scales linearly with molecular size, the computational cost may increase dramatically with the basis set size. Therefore reducing the prefactors remains an important issue. The above mentioned Orbital Minimization (OM) and Optimal Basis Density-Matrix Minimization methods suffer from ill-conditioning problems and therefore require frequently an excessive number of iterations, while in both OM and FOP methods the pre-knowledge about the bonding properties is needed to form the initial guess [6]. Finally most of the linear scaling methods are known to fail when applied to metallic systems. Still, in spite of fact that the various linear scaling methods and algorithms have been developed and applied in a large variety of fields, however, very little is known on the performance of such approaches for the computation of the NLO properties. Apart from the above-mentioned techniques, a branch of methods that explicitly divide molecular systems into smaller fragments has been proposed, from which divide and conquer [14–17], fragment molecular orbital [18–25], and density matrix renormalization group methods [26–28] are probably the most well known at the ab initio HF level of theory and are most actively being developed.

Our treatment is different from other linear scaling methods in adopting the concept of the Region Localized Molecular Orbitals (RLMOs) [29] and it has recently been extended to be applicable for three-dimensional (3D) systems [30]. As an approach toward post-Hartree–Fock, we developed the elongation local MP2 (ELG-LMP2) method [4]. We also confirmed that the RLMO is very useful to evaluate local excitation around chromophore center in a large system at single excitation configuration interaction (CIS) and TDHF levels [31]. Furthermore, a so-called "cut-off" technique for the elongation method makes possible to deal with huge number of 2e-integrals of gigantic systems [32]. Additionally Quantum Fast Multipole Method (QFMM) method was incorporated into the elongation method at the Hartree-Fock and Density Functional Theory (DFT) levels [33–35].

In last decade, the elongation finite-field (elongation-FF) method has been developed in the theoretical chemistry laboratory [36–42]. In this approach, molecular (hyper)polarizabilities are obtained by numerically differentiating the total energy with respect to electric fields. This method has been applied to many quasi-one-dimensional π-electron conjugated organic polymers, such as polyacetylene, polydiacetylene, polythiopene, and their derivatives at both levels of semi-empirical and ab initio method. Also recently it has been applied to gigantic systems as BNC nanotubes [40] and a porphyrin wire [41]. However, we found the fact that the fourth order energy derivatives with respect to electric field sometimes cause a big error around 10–20 % of conventional values especially in second hyperpolarizability (γ) even with 10^{-6} a.u./atom difference in the total energy between those by conventional method and elongation method.

The purpose of this work is to improve the accuracy of our elongation method to evaluate non-linear optics (NLO) properties of polymers and to develop this method so as to be able to analyze the relationship between structure and NLO properties. Therefore, more accurate total energy evaluations with total energy error less than 10^{-10} a.u./atom is strongly desired to get acceptable hyper-polarizability values. For this purpose, we have extended the traditional elongation method which allows one to get very accurate total energies so that one can get highly accurate higher order derivatives of the total energy efficiently. By this technique, the physical properties for strongly delocalized systems like carbon nanotubes and photonic polymers can be obtained under various applied external fields. The calculation of the NLO of such systems will play an important role in the next generation information and computing industry that will require device designs at the nano-level. In this book, we show how total energies have been improved by our new development and present some preliminary NLO property results. This method development is essential for the elongation method and its applications to be able to treat bio-/nano-systems with orbital delocalization and its functional designing with high accuracy. This method makes it possible to treat huge strongly delocalized systems that are formidable by the conventional method. In this chapter, we focus not only on the total energy improvement but also on NLO property calculations. We have already reported computational time of elongation method in other articles (see, for example, [4, 31–35, 40]).

6.1 Outline of Elongation Method

The elongation method has been implemented in the GAMESS program package [42]. In the elongation procedure, the delocalized canonical molecular orbitals (CMOs) of a starting cluster are first localized into frozen and active regions in the specified parts of the molecule. The "specified parts" means appropriate size of A region and B region to get good localization as well as high accuracy for total energies (error of less than 10^{-8} a.u./atom) in comparison with that obtained by conventional methods. For getting this criterion, we normally select one unit as the A region and 3–5 units as the B region on which active RLMOs will be formed for the interaction with attacking monomer. Next, a monomer is attacking to the active region of the cluster, and the eigenvalue problem is solved by disregarding the localized molecular orbitals (LMOs) which have no or very weak interaction with the attacking monomer. By repeating this procedure, the length of the polymer chain is increased step by step

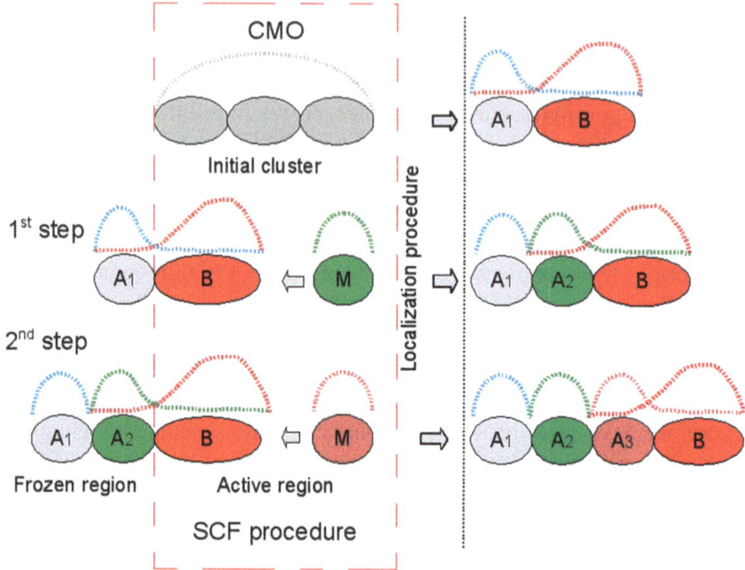

Fig. 6.1 The graphical scheme of elongation method

to any desired length. The obvious advantage of the approach lies in the fact that one can avoid solving very large secular equations for large aperiodic systems (Fig. 6.1).

Since the detailed explanation in making region LMO (RLMOs) in an efficient manner has been described in other articles by Gu et al. [29], here only the essential points of this method are explained with the illustrations as shown in Fig. 6.2. One can partition the starting cluster into two regions, region A (frozen region) and region B (active region) and localize the CMOs into these two regions. Region B is the one with atoms adjacent to the interactive end of the cluster whereas region A is at the opposite end far away from the interactive center. It should be mentioned here that by using the density matrix, the partition of the starting cluster into two regions is unique since the AOs belonging either to A or B region are well defined. This is different from the other localization schemes, where the partitions are not unique. As one has already seen the division of CMOs into two regions is not so straightforward and poor selection leads to poor localization. The desired RLMOs can be obtained according to five steps as shown below and in Fig. 6.2.

1. As shown in Fig. 6.2 (step1), the density matrix in the orthogonal atomic orbital (OAO) basis is transferred from AO basis by X, where the following idempotence relation of the density matrix in an OAO basis

$$D^{OAO}D^{OAO} = 2D^{OAO} \tag{6.1}$$

is proved and the eigenvalues of D^{OAO} must be either 2 or 0. Therefore, the eigenvectors of are well separated into occupied or vacant subspaces.

2. A regional orbital (RO) space is constructed by separately diagonalizing the $D^{OAO}(\mathrm{A})$ and $D^{OAO}(\mathrm{B})$, where$D^{OAO}(\mathrm{A})$ and $D^{OAO}(\mathrm{B})$ are the sub-blocks of D^{OAO} with AOs belonging to A and B regions, respectively. These eigenvectors span the RO space. The transformation from OAOs to ROs is given by the direct sum of T^A and T^B, where T^A and T^B are the eigenvectors of $D^{OAO}(\mathrm{A})$ and $D^{OAO}(\mathrm{B})$, respectively. Figure 6.2 (step 2) shows the diagonalization of the matrices of $D^{OAO}(\mathrm{A})$ and $D^{OAO}(\mathrm{B})$ and the schematic construction of the T matrix. The corresponding eigenvalues are divided into three sets corresponding to ROs that are approximately doubly-occupied (the value is close to 2), singly-occupied (close to 1) and empty (close to 0). The singly-occupied orbitals in A and B can be used to construct hybrid orbitals to form covalent bonding/antibonding pairs. For a non-bonded system, such as water chain, there are only two sets, either doubly-occupied or empty. Figure 6.2 (step 2) shows the latter case.
3. The RO density matrix is obtained by transforming the D^{OAO} by T matrix combined by T^A and T^B as shown in Fig. 6.2 (step 3). Using Eq. (6.1) and the unitary condition $TT^{\dagger} = T^{\dagger}T = 1$, one can verify that

$$D^{RO}D^{RO} = 2D^{RO} \tag{6.2}$$

4. Except for orthogonalization tails the ROs given above are completely localized to region A or region B. However, they are not completely occupied or unoccupied. Thus, final step is to carry out a unitary transformation between the occupied and unoccupied blocks of D^{RO} to keep the localization as large as possible. Then the only non-zero elements of D^{RLMO} are equal to 2 (cf. Eq. 6.2). As shown in the Fig. 6.2 (step 4) when D^{RO} is transferred back into D^{OAO}, one can see that $T\zeta$ gives the transformation matrix of D^{OAO}. The $T\zeta$ values have plenty of freedom in the framework of a unitary transformation because the eigenvalues of D^{OAO} are degenerate by 2 in occupied space and by 0 in unoccupied space. Nevertheless, we can define the transformation matrix, $T\zeta$, so that the RLMOs are localized to the largest extent possible into A(frozen) or B(active) regions "through T matrix", because we can get the RLMO coefficients similar to the T matrix in this special localization scheme. Thus, finally obtained $T\zeta$ becomes very similar to the original T matrix in OAO basis.
5. Finally, the unitary transformation from CMO to RLMO is given in the Fig. 6.2 (step 5). After transferring $T\zeta$ back to the AO basis by using X, we can get the RLMO coefficients in the AO basis as shown in the right hand side of the Fig. 6.2 (step 5).

After the well-localized RLMOs are obtained by the above procedure, the Fock matrix based on the RLMOs is expressed only by using active RLMOs as shown in Fig. 6.3, in which frozen RLMOs are removed before interaction with the attacking monomer. Then our working space is defined by the RLMOs assigned to the B region together with the CMOs of the attacking monomer. For a given suitably large starting cluster, the interaction between the frozen region and the attacking monomer

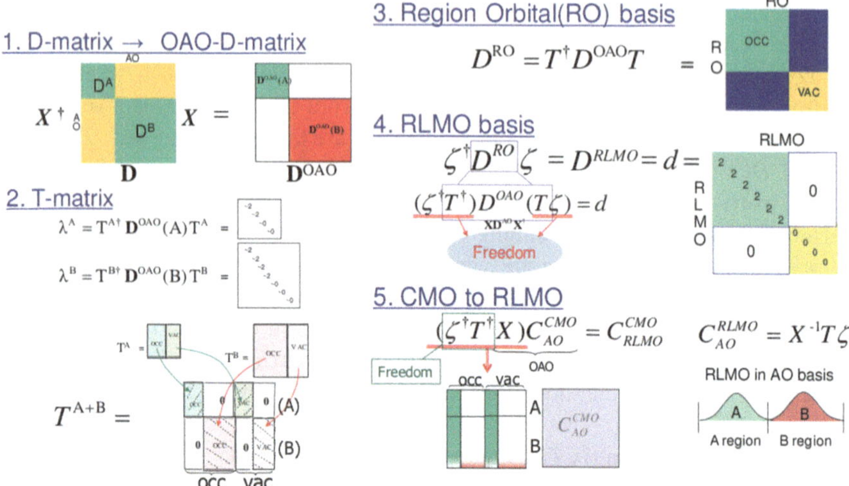

Fig. 6.2 Schematic illustration to show how to make RLMOs

is minimized by using RLMOs. In our treatment, the proper size of the starting cluster is automatically detected during elongation from a rather small starting cluster and the elongation process with a reduced number of RLMOs is initiated after the condition to remove the frozen orbitals is satisfied. The elongation Hartree-Fock equation is solved self-consistently in the localized orbital space of the interactive region, or more precisely, the working space consists of RLMOs of the active region and CMOs of the attacking monomer. This solution yields a set of CMOs in the reduced space which can be localized again into a new frozen region and a new active region. The whole procedure is repeated until the desired length is reached. The important feature of the elongation method is that the Hartree-Fock equation is solved only for the interactive region instead of the whole system. As the system enlarges, the size of the interactive region is almost the same as that of the starting cluster and the CPU required in the elongation SCF is more or less constant. Additionally, the "Cut region" in the AO based Fock matrix of Fig. 6.3, the region far from the attacking monomer, has no contribution on the final RLMO based Fock matrix because the corresponding coefficients part of the active RLMOs should be almost zero as written by "∼0" in the active RLMO coefficients. So we can get rid of not only the frozen RLMOs, but also the frozen part in the AO based Fock matrix elements, leading to big reduction in making Fock matrix and O(N) computational time during the elongation process. The details about cutoff procedure are described in our recent papers [32, 35].

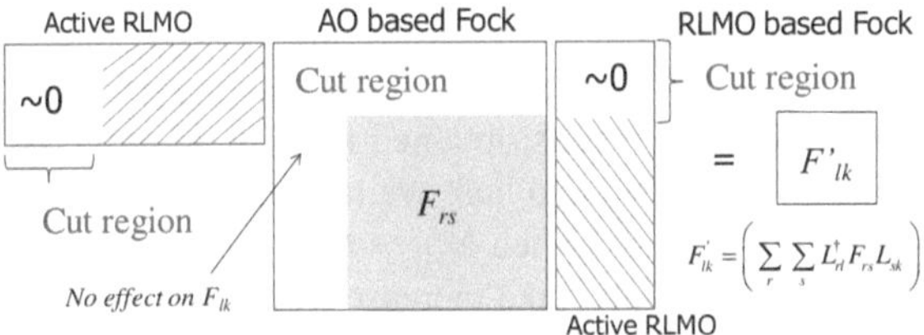

Fig. 6.3 Schematic illustration to make RLMO based Fock matrix

6.2 Elongation Method Deficiency

Although we can elongate both periodic and aperiodic one-dimensional systems, some problems remain in strongly delocalized systems. For example, if the system has strongly delocalized π orbitals, our treatment does not provide satisfied accuracy in the total energies in comparison with conventional direct calculations. This causes sometimes a big error especially in the second hyperpolarizability (γ) which is given by fourth order energy derivatives. However, significantly large hyperpolarizability is mostly found in strongly delocalized large systems and thus a highly accurate computational method for such systems is desired to design useful NLO materials.

The example which shows the difficulty to calculate accurate energies for delocalized systems with the traditional elongation method is given by the carbon-rich single-wall boron nitride/carbon nanotube (SW-BN/CNT) as shown in Table 6.3. It is reported that the electronic band gap of single-wall boron nitride nanotubes (SW-BNNTs) can be reduced by adding C atoms into the structure [33, 34] and so carbon-rich BN/CNT that possesses delocalized π-electrons can be a good model to examine the applicability of our new treatment. Calculations of the electronic structures were performed for pristine SW-BNNTs with different diameters and chirality. We confirmed that the accuracy depends on the size of the diameter rather than on the basis set. The errors in total energy become larger with increasing diameter size, because π-electron delocalization effects become stronger in larger diameter nanotubes [40]. Then we focus on (4, 4) BN/C heterostructured nanotubes ($(BN)_xC_y$ $(x,\ y = 1 - 4,\ x + y = 5)$) including carbon atoms that make the delocalization stronger with included C ratio in the framework of HF/STO-3G theory. The periodic structures are optimized by using the Gaussian 03 program [46] at the B3LYP/6-31G level. The total energy difference between elongation method and conventional direct method is listed in Table 6.3, where some of data for different sizes are from the Ref. [46]. The N_{st} in the table means the starting cluster size in our elongation method. In Table 6.3, the Number of units 6 and 7 (112 and 128 atoms, respectively) corresponds to the number of units in the starting cluster and in the first elongation step for $N_{st} = 6$, respectively, while the Number of units 10 and 11 (176 and 192 atoms, respectively) corresponds to the number of units in the starting cluster and in the first elongation step for $N_{st} = 10$, respectively. This information is important to see how large errors are induced by one elongation step. In addition, the information concerning the last step, with the Number of units 20 (336 atoms) is also important. We also show the results obtained during the elongation procedure for 11 units and 17 units as examples. The results obtained for 6 to 20 units have already been reported in Table 6.6 of Ref. [40], and here we just present a small cross section of that data to document the results. When $N_{st} = 6$, it can be seen that the errors per atom for four systems are very small, but increase with the carbon content because of more delocalization in carbon's nature. Especially for carbon rich system, $(BN)_1C_4$, the error $\sim 10^{-6}$ a.u./atom is not acceptable to calculate reliable γ as seen from the example of metalloporphyrin arrays. So, for $(BN)_1C_4$ and $(BN)_2C_3$, we recalculated their energies by using the larger sized starting cluster $N_{st} = 10$. As a result, the error per atom is decreased to 10–8 a.u. for $N_{st} = 10$, but β and γ values still show unstable curves as a function of the number of units as seen in Fig. 3 of Ref. [40]. We have already confirmed that the delocalized π-orbitals from carbon atoms give small band gaps, which cause insufficient accuracy in the total energy calculations during elongation process which is required for the calculation of nonlinear optical properties, for example, second (hyper)polarizabilities. To solve the problem of insufficient accuracy and inefficiency in delocalized systems we propose here a simple technique to incorporate the delocalized frozen orbitals into the active space, and the results tested for some nano- and bio-systems using this treatment are shown in the next section.

6.3 Orbital Basis Concept

The problem in the accuracy is caused by the fact that we only treated the RLMOs as the "region" basis when we judge if we discard the frozen RLMOs that located farthest from attacking monomer or not. That is, when the contribution from active RLMOs on the frozen terminal unit

$$P(A_1) = \sum_{r}^{on A_1} \sum_{s}^{on A_1} C_{ri}^{ActiveRLMO} S_{rs} C_{si}^{ActiveRLMO} \tag{6.3}$$

is less than the threshold value (for example, 10−5), we can start the elongation and then we can disregard all the frozen RLMOs on the frozen terminal unit. The active RLMOs are already sufficiently included in the interaction with attacking monomer under the defined threshold. If the threshold value, $P(A_1)$, is set tight, the elongation starts late after several elongation steps, and then the number of active RLMOs to be included into active space increases, leading more accurate results with longer CPU time. On the other hand, the threshold value is set too loose, the elongation starts early, and then the number of active RLMOs decreases, leading to less accurate results with shorter CPU time. For delocalized systems, however, the elongation process sometimes even does not start unless we set the threshold very loose because some orbitals that cannot inherently be localized still remain on the frozen region. This fact forces the threshold very loose to start elongation process for delocalized systems and thus the obtained accuracy becomes worse compared to that for other systems. We call it as "region basis concept" because a set of the frozen orbitals in the frozen region has two choices of either being included into active region or discarded. However, the number of delocalized RLMOs (almost similar shape to CMOs) is limited to a few frozen orbitals and kept almost constant during the elongation process. So, CPU time consumed for this inclusion may not be accumulated during the process. To improve the accuracy in "region basis concept" mentioned above, we introduce "orbital basis concept" together with the "region basis concept" for the more effective selection of necessary active orbitals.

For any RLMO of the starting cluster, one can initially define the overlap between pair of frozen RLMO and active RLMO as,

$$Q_{ij} = \langle \phi_i^{FrozenRLMO}(A) | \phi_j^{ActiveRLMO}(B) \rangle \tag{6.4}$$

With this quantity we can find the ith frozen RLMOs that is associated with jth active RLMOs with an overlaps larger than the threshold value (we set 10^{-4} as the default). If some frozen RLMOs have values of Q_{ij} larger than the threshold, these orbitals are still assigned to the delocalized group even after the localization process and are transferred back into active RLMO group to join the interaction with the attacking monomer. Only the remaining frozen RLMOs are assigned to "truly frozen" orbitals that never interact with the attacking monomer hereafter. In the next elongation step, we also check the Q_{ij} value again in the next boundary between frozen region and active regions, and then detected delocalized orbitals are always transferred back into active RLMO group. By continuing this treatment through all the elongation steps, delocalized orbitals are always picked up and shifted to active space to give a more correct result. In addition, the number of shifted orbitals is not increased at the next elongation step because the number of targeted orbitals on the boundary between frozen and active regions remains always essentially the same. As a result, the computational time is not accumulated and even become shorter since this treatment makes the elongation start earlier from small starting cluster with keeping the accuracy high. Finally we get excellent agreement with the total energy calculated by conventional method even for strongly delocalized systems. The schematic illustration of delocalized orbitals is drawn in Fig. 6.4 as a model of SW BN/CNT. It can be seen from the figure that the blue orbitals originally assigned as frozen RLMOs have some overlap with pink active RLMOs over the A and B regions as far as the tailing of frozen RLMOs is permeated into B region. These orbitals correspond to those we should still include into the active RLMO group towards the interaction with attacking monomer.

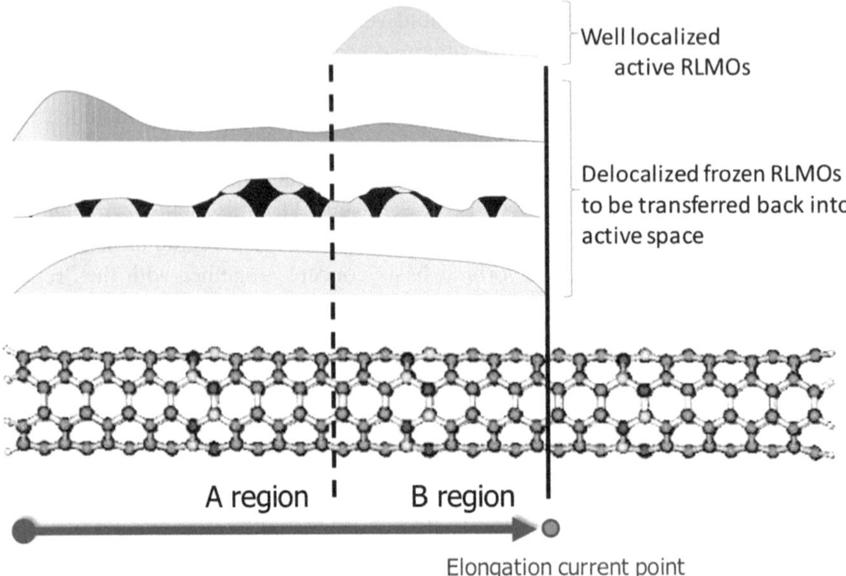

Fig. 6.4 Schematic illustration of the shape of well localized (ideal) active RLMOs and delocalized frozen RLMOs to be shifted to active RLMO group

6.4 Eigenvalue Problem

At every elongation step the SCF equations are solved only in small subspace (B+M) of full variational space, regardless of the total number of atoms in system. The clear advantage of EM is that the eigenvalue problem is solved for active region only. However evaluation of some electronic properties requires the eigenvalue solution for the entire system. Obtaining eigenvectors and eigenvalues for the whole molecular chain would demand time-consuming diagonalization of full Fock matrix.

In the present work we try to minimize the cost of full diagonalization by employing prediagonalized energy matrix based on regional LMOs.

It should be stressed that EM itself does not work directly with CMOs, but proposed in this section algorithm allows to recover from this condition and to obtain CMOs omitting numerous iterative diagonalization procedures. And it is of great importance to have the CMOs to be able to describe delocalized phenomena, like band structure, low frequency vibrational modes such as lattice vibration, Wannier type excitons etc.

Computational Details

The elongation method with the regional localization scheme as it is implemented in GAMESS program package has been used for *ab initio* treatment with STO-3G basis set. The optimization calculation of structures, used in CPU check, has been done by means of Møller–Plesset perturbation theory with 6-31G(d) basis set. The SCF density convergence criteria has been set up to 10^{-8}.

Methodology

In order to obtain molecular orbital energies and CMOs of the whole system within EM and to avoid the diagonalization of the full Fock matrix we adopted the following scheme:

$$E(A+B+M) = C_{AO}^{CMO\dagger}(A+B+M)F_{AO}^{AO}C_{AO}^{CMO}(A+B+M) \tag{6.5}$$

where unknown moiety C_{AO}^{CMO}(A+B+M) can be obtained from

$$C_{AO}^{CMO}(A+B+M) = C_{AO}^{LMO}(A+B+M)X^{\dagger}(A+B+M) \tag{6.6}$$

where C_{AO}^{LMO}(A+B+M) is the total LMO matrix which consists of separate LMOs associated with A (frozen region) produced from every elongation step and LMOs of active region from the final step:

$$C_{AO}^{LMO}(A+B+M) = C_{AO}^{LMO}(A)_1 \cdots \oplus C_{AO}^{LMO}(A)_{n-1} \oplus C_{AO}^{LMO}(B+M)_n \tag{6.7}$$

and X is the direct sum of the constituent YYs, where each YY describes unitary transformation from LMO to CMO at every elongation step:

$$YY = C_{AO}^{CMO\dagger} S C_{AO}^{LMO} \tag{6.8}$$

where S is the overlap matrix in AO basis.

Results and Discussion

The obtained energy matrix for linear hydrogen chain $(H_2)_{12}$ is presented in Table 6.1. As one can see this matrix is almost diagonal, though we did not carried out any diagonalization procedures. However there is some admixture of non-zero off-diagonal elements. This stems from the small 'tailing' effect of frozen A region onto active

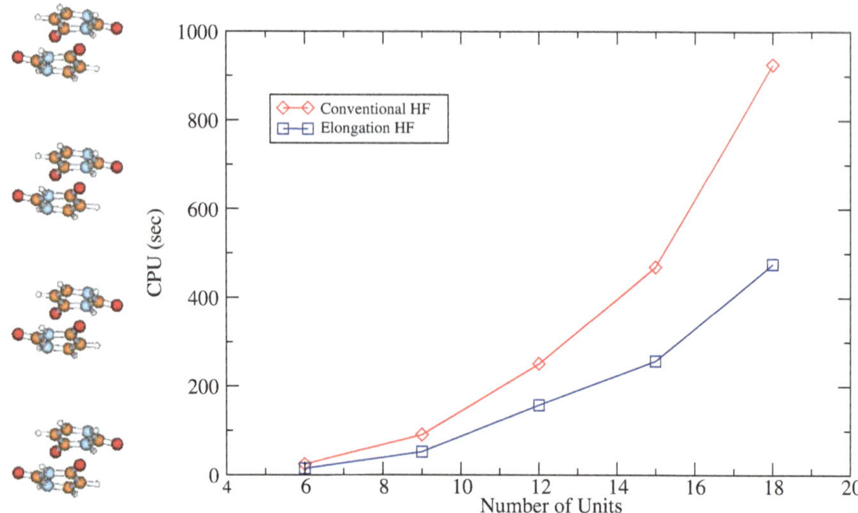

Fig. 6.5 The structure of investigated molecular chain. CPU time usage in diagonalization procedure. Each unit corresponds to the one uracil pair

region. As the consequence, the presence of non-zero off-diagonal elements affects the accuracy of diagonal eigenvalues.

The calculated orbital energies from elongation HF and from conventional HF are shown in Table 6.2.

The largest discrepancy among occupied MO energies is $6*10^{-4}$ and among vacant MO energies even worse $4*10^{-3}$. To eliminate off-diagonal elements we performed rotation of our energy matrix using Jacobi algorithm. After few rotations the MO energy values improved significantly and the gap between conventional and elongation values decreased in 3 orders of magnitude. The corresponding orbital energies after rotation are tabulated in Table 6.2. The density matrix derived from elongation eigenvectors matches up to the forth decimal the density matrix of conventional MOs.

Time used for collecting all LMOs, YYs and constructing energy matrix is quite insignificant in comparison to rate-determining rotation step. Therefore we focused on this bottleneck of eigenvalue solution. We performed checking of CPU time usage for diagonalization of our energy matrix and compared it to the CPU time used for diagonalization of Fock matrix in conventional calculation. For this purpose we used stacked uracil pares optimized at MP2 level of theory. The optimized distance between two uracil planes, constituting the elementary unit, was 3 Å and the distance between units has been arbitrary chosen to be 6 Å. The results are shown on Fig. 6.5.

Table 6.1 Energy matrix of $(H_2)_{12}$

−0.5836	−0.0009	0.0006	0.0000	0.0000	0.0000	−0.0006	−0.0007	0.0004	0.0000	0.0000	0.0000
−0.009	**−0.5805**	−0.0004	0.0000	0.0000	0.0000	0.0000	0.0000	0.0000	0.0000	0.0000	0.0000
0.0006	−0.0004	**−0.5818**	0.0000	0.0000	0.0000	−0.0005	−0.0006	0.0004	0.0000	0.0000	0.0000
0.0000	0.0000	0.0000	**0.6641**	0.0051	0.0048	0.0000	0.0000	0.0000	0.0034	−0.0047	−0.0033
0.0000	0.0000	0.0000	0.0051	**0.6848**	0.0045	0.0000	0.0000	0.0000	0.0000	0.0000	0.0000
0.0000	0.0000	0.0000	0.0048	0.0045	**0.6771**	0.0000	0.0000	0.0000	0.0033	−0.0046	−0.0033
−0.0006	0.0000	−0.0005	0.0000	0.0000	0.0000	**−0.5840**	0.0000	0.0000	−0.0002	−0.0001	0.0000
−0.0007	0.0000	−0.0006	0.0000	0.0000	0.0000	0.0000	**−0.5819**	0.0000	0.0002	0.0000	0.0001
0.0004	0.0000	0.0004	0.0000	0.0000	0.0000	0.0000	0.0000	**−0.5800**	0.0000	−0.0001	0.0002
0.0000	0.0000	0.0000	0.0034	0.0000	0.0033	−0.0002	0.0002	0.0000	**0.6619**	0.0000	0.0000
0.0000	0.0000	0.0000	−0.0047	0.0000	−0.0046	−0.0001	0.0000	−0.0001	0.0000	**0.6754**	0.0000
0.0000	0.0000	0.0000	−0.0033	0.0000	−0.0033	0.0000	0.0001	0.0002	0.0000	0.0000	**0.6887**

Table 6.2 MO energies of $(H_2)_{12}$ in a.u.

MO	1	2	3	4	5	6	7	8	9	10	11	12
Elg	−0.583972	−0.583610	−0.581893	−0.581843	−0.580459	−0.580046	0.661925	0.664051	0.675390	0.677101	0.684834	0.688675
Elg*	−0.584593	−0.583675	−0.582471	−0.581231	−0.580199	−0.579655	0.658212	0.663549	0.671223	0.679607	0.687148	0.692237
Conv	−0.584592	−0.583675	−0.582469	−0.581230	−0.580199	−0.579655	0.658211	0.663549	0.671226	0.679611	0.687148	0.692235

Elg*—the eigenvalues after rotation of energy matrix

6.5 Applications

It should be noted that both methods have the qualitative agreement in produced orbital energies, but upon the increasing size of the studied system the diagonalization time of energy matrix is two times less than diagonalization of Fock matrix in conventional HF method. This gain in time becomes possible because of preoptimized shape of energy matrix. As it can be seen from the Table 6.1 the energy matrix is essentially diagonal, so it takes less time to diagonalize such sparse matrix than to do the diagonalization from the scratch as in the case of conventional HF method.

Conclusions

The algorithm for computing MO energies has been designed for Elongation Method. The obtained results yield high accuracy in comparison with the conventional approach. The efficient CPU time usage has been achieved for the diagonalization procedure.

6.5.1 SW-BN/CNT

The treatment described in the previous section was applied first to the carbon rich (4,4) SW- BN/CNT, $(BN)_1C_4$, system that provided worse results in accuracy compared to $(BN)_5$, $(BN)_4C_1$, $(BN)_3C_2$, and $(BN)_2C_3$ as shown in Table 6.3. The resultant errors by the revised elongation method, denoted by New_Elg, are listed in Table 6.4 in comparison with those by traditional elongation method, denoted by Old_Elg. The table shows that the accuracy was around two orders of magnitude improved by the revised elongation method. These results are promising to get reliable results when we apply it to more delocalized systems not only in conjugated hydrocarbon systems but also metallic systems.

Table 6.3 Structure of the (4, 4) pristine and BN/C heterostructured nanotubes (BN)xCy (x, y = 1 − 4, x + y = 5)

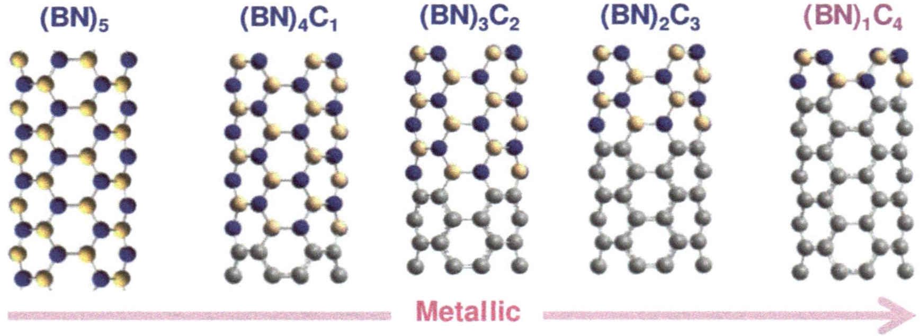

Number of units	Number of atoms	$N_{st} = 6$ (Starting cluster size)				$N_{st} = 10$	
		$(BN)_4C_1$	$(BN)_3C_2$	$(BN)_2C_3$	$(BN)_1C_4$	$(BN)_2C_3$	$(BN)_1C_4$
6	112	0.000E + 00	0.000E + 00	0.000E + 00	0.000E + 00		
7	128	7.945E − 10	3.680E − 09	2.336E − 08	3.059E − 07		
10	176	7.379E − 09	4.901E − 08	5.006E − 07	6.209E − 06	0.000E + 00	0.000E + 00
11	192	7.748E − 09	3.927E − 08	4.255E − 07	5.137E − 06	3.167E − 10	9.695E − 09
17	288	1.144E − 08	5.598E − 08	5.851E − 07	7.054E − 06	1.872E − 09	2.881E − 08
20	336	1.505E − 08	7.742E − 08	7.713E − 07	9.186E − 06	5.283E − 09	3.829E − 08

The elongation errors (a.u./atom) in total energy are shown for different starting cluster size (N_{st})

Table 6.4 Error per atom (in a.u.) introduced by the old and new elongation methods for (4,4)SW-BN/CNT, $(BN)_1C_4$, at the HF/STO-3G level

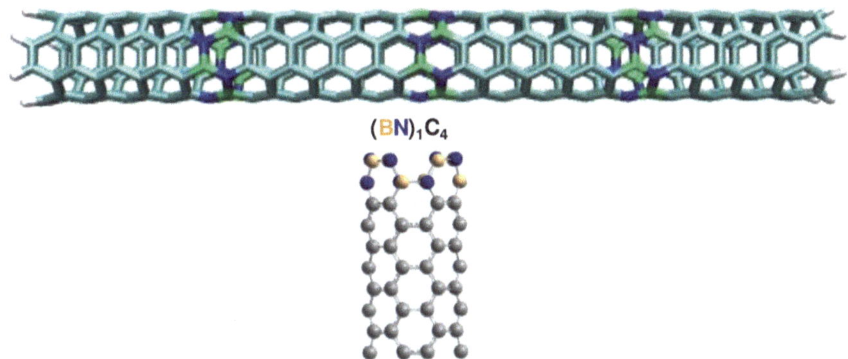

Atoms	Total energy Conv (in a.u.)	Old_Elg Error/atom	New_Elg Error/atom
112	−3654.67279190	0.00E + 00	0.00E + 00
128	−4253.10068386	3.06E − 07	1.68E − 09
144	−4851.70684615	2.97E − 06	8.93E − 09
160	−5450.18308411	4.60E − 06	3.66E − 08
176	−6048.76373323	6.21E − 06	1.75E − 08
192	−6674.65602795	5.14E − 06	1.50E − 08
208	−7273.24074195	5.71E − 06	1.70E − 08

6.5.2 Polyacene

Another delocalized example, polyacene, as shown in Table 6.5, is tested by using both old and new elongation methods as a first step to apply our method to the most intractable two dimensional graphene sheet. This system possesses even more delocalized π-orbitals than SW-BN/CNT because all of them locate completely perpendicular to the plane. This fact can be suggested also from larger errors in SW-BN/CNT system with larger diameter [40]. With the old elongation method, the accuracy depends on the basis set size because larger basis set makes the localization worse, leading to larger errors in total energy. However, one can see that the errors by STO-3G basis set are improved only by one-order of magnitude in the new method while those by 6-31G(d) are by two-orders of magnitude. This fact suggests that this correction technique provides always good results even for strongly π-electron delocalized systems almost regardless of the basis set size. If the more tight threshold for Q_{ij} is used, the more accurate results must be expected because a few more frozen orbitals accompanied by small overlap with active RLMOs are additionally joined to the interaction with attacking monomer. If an infinite zero threshold is used, all the

Table 6.5 Error per atom (in a.u.) introduced by the old and new elongation methods for polyacene chain at the HF/STO-3G and 6-31G(d) level

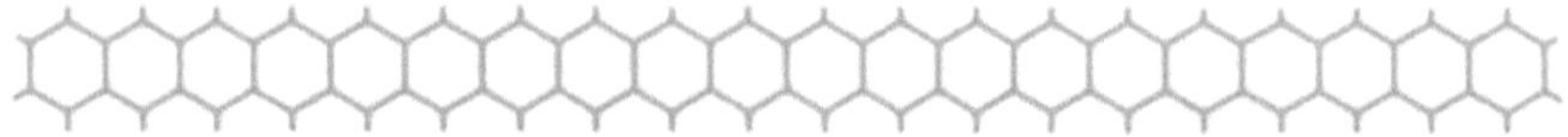

Step	Number of units(rings)	Number of atoms	STO-3G		6-31G(d)	
			Old_Elg Error/atom	New _Elg Error/atom	Old_Elg Error/atom	New_Elg Error/atom
0	6	42	0.00E + 00	0.00E + 00	0.00E + 00	0.00E + 00
1	7	48	1.51E – 05	2.51E – 07	1.03E – 04	2.06E – 07
2	8	54	2.97E – 05	5.64E – 07	1.64E – 04	5.13E – 07
3	9	60	4.24E – 05	8.61E – 074	2.42E – 04	8.52E – 07
4	10	66	5.31E – 05	1.12E – 06	2.83E – 04	1.17E – 06
5	11	72	6.21E – 05	1.34E – 06	3.17E – 04	1.46E – 06
6	12	78	6.98E – 05	1.53E – 06	3.46E – 04	1.73E – 06
7	13	84	7.63E – 05	1.69E – 06	3.71E – 04	1.96E – 06
8	14	90	8.20E – 05	1.83E – 06	3.93E – 04	2.17E – 06
9	15	96	8.70E – 05	1.95E – 06	4.12E – 04	2.35E – 06
10	16	102	9.14E – 05	2.06E – 06	4.28E – 04	2.53E – 06
11	17	108	9.53E – 05	2.15E – 06	4.58E – 04	2.68E – 06
12	18	114	9.87E – 05	2.24E – 06	4.84E – 04	2.84E – 06
13	20	126	1.01E – 05	2.32E – 06	4.77E – 04	3.00E – 06

orbitals are admitted to the SCF calculations and then obtained results are completely same as those by the conventional method though this situation is meaningless.

6.5.3 *β-carotene*

As an application to bio-system, we applied our treatment to β-carotene as shown in the top of Table 6.6. The two basis sets of STO-3G and 6-31g(d) were also implemented for the geometry optimized at HF/6-31G level to check the accuracy from starting cluster size of 6 units. The STO-3G results show around two-order improvement in the total energy by the new elongation method, while the 6-31G(d) results show around three-order improvement as seen in the case of polyacene. The errors caused by traditional elongation method become larger with basis sets size as seen from the table, while new treatment presented here is not so sensitive on the basis set used.

Table 6.6 Error per atom (in a.u.) introduced by the old and new elongation methods for β-carotene at the HF/STO-3G and 6-31G(d) level

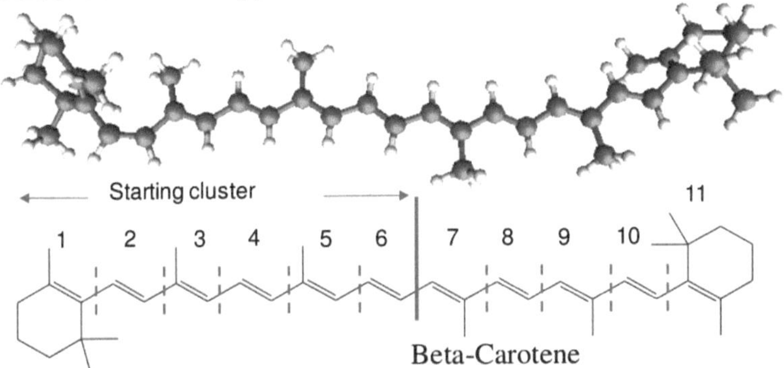

Step	Number of units	Number of atoms	STO-3G Old_Elg Error/atom	STO-3G New_Elg Error/atom	6-31G(d) Old_Elg Error/atom	6-31G(d) New_Elg Error/atom
0	6	51	0.00E + 00	0.00E + 00	0.00E + 00	0.00E + 00
1	7	58	1.01E − 07	7.43E − 10	1.10E − 06	3.31E − 10
2	8	62	3.18E − 07	8.94E − 10	1.71E − 06	1.60E − 09
3	9	69	5.76E − 07	1.88E − 09	2.49E − 06	1.42E − 09
4	10	73	7.86E − 07	2.90E − 09	2.95E − 06	2.47E − 09
5	11	96	7.39E − 07	6.07E − 09	3.57E − 06	2.47E − 09

6.5.4 Lycopene

We applied the revised elongation method to another bio-system to confirm the applicability to yet another system. As an example for a rather small but delocalized systems, our treatment was tested on the lycopene oligomer as shown in Table 6.7. Here we examined more details related to the B-region size dependency. The geometry was adopted from experimental data [47] but only hydrogens were optimized at the B3LYP/6-31G level. As for B-region = 3 units, the elongation started just after the starting cluster calculation of 5th (A-region = 2 and B-region = 3) since the definition of Eq. 6.3 to remove frozen orbitals was already smaller than the threshold at 5th units. The accuracy was drastically improved by four-orders of magnitude using the new elongation method. The N_{shift} denotes the number of frozen orbitals that were shifted to the active space. One can see that the number of the shifted orbitals is almost constant during the elongation. Therefore the total number of active RLMOs does not accumulate with chain length even after the orbitals are shifted and thus the required CPU times at each step is almost kept constant during the elongation process. As for B-region = 4 units, the elongation started also just after the starting cluster calculation of 6th units and the N_{shift} shows the same behavior as the case of B-region = 3 units. As for B-region = 5 units, similarly, the condition to remove

Table 6.7 Error per atom (in a.u.) introduced by the old and new elongation methods for lycopene chain for different B region sizes at the HF/3-21G level

A region=2

Unit Numbering 1 2 3 4 5 6 7 8 9 10 11 12 13 14 15

(Number of atoms in a unit) (7) (4) (7) (4) (7) (4) (7) (4) (7) (6) (10+1)

Number of atoms in current step 25 29 36 40 47 51 58 62 69 73 80 86 96

Number	Number	B region = 3 units		N_{shift}	B region = 4 units		N_{shift}	B region = 5 units		N_{shift}
of units	of atoms	Old_Elg Error/atom	New_Elg Error/atom		Old_Elg Error/atom	New_Elg Error/atom		Old_Elg Error/atom	New_Elg Error/atom	
5	36	0.00E + 00	0.00E + 00	7						
6	40	2.65E – 05	−3.84E – 09	8	0.00E + 00	0.00E + 00	7			
7	47	5.23E – 05	2.16E – 09	8	1.05E – 06	−2.23E – 10	8	0.00E + 00	0.00E + 00	7
8	51	6.75E – 05	7.22E – 09	7	3.38E – 06	−3.69E – 10	8	1.58E – 007	6.08E – 011	8
9	58	8.69E – 05	1.06E – 08	8	4.77E – 06	8.59E – 10	7	4.96E – 007	3.66E – 010	8
10	62	9.17E – 05	1.80E – 08	8	6.78E – 06	6.77E – 09	8	8.36E – 007	1.27E – 009	7
11	69	1.09E – 04	1.62E – 08	8	8.41E – 06	4.59E – 09	8	1.18E – 006	1.17E – 009	8
12	73	1.18E – 04	2.42E – 08	10	1.04E – 05	1.12E – 08	8	1.50E – 006	2.30E – 009	8
13	80	1.20E – 04	2.26E – 08	7	1.08E – 05	1.02E – 08	10	1.69E – 006	2.17E – 009	8
14	86	1.27E – 04	3.28E – 08	10	1.11E – 05	9.61E – 09	7	1.77E – 006	4.74E – 009	10
15	96	1.14E – 04	2.58E – 08	–	9.88E – 06	7.93E – 09	–	1.60E – 006	3.26E – 009	–
			Sumof shifted orbitals	81		Sumof shifted orbitals	71		Sumof shifted orbitals	64

Table 6.8 Error per atom (in a.u.) introduced by the new elongation method for lycopene chain with B region = 2 and 5 units at the HF/6-31+G(d,p) level

Number of units	Number of atoms	New_Elg Error/atom B region = 2 units	B region = 5 units
5	36	0.00E + 00	
6	40	0.00E + 00	
7	47	0.00E + 00	0.00E + 00
8	51	1.37E − 11	9.22E − 11
9	58	1.21E − 10	1.09E − 10
10	62	1.15E − 10	1.23E − 10

frozen orbitals is satisfied in the starting cluster of 7th units. The errors caused by the new elongation method are almost the same $\sim 10^{-8\sim 9}$ a.u./atom for all the three different B-region sizes. The N_{shift} coincided with one step later with increasing one B-region unit, because localization between A and B regions starts one step later. For example, $N_{shift} = 7$ for all different B-region sizes in the initial step means the number of shifted orbitals in the boundary 1st unit and 2nd unit just after starting cluster calculation toward 1st elongation step. But these orbitals are frozen in the next elongation step and then 8 orbitals were newly selected for shifting to the active space in the second elongation step. Finally, we can conclude that the new elongation method detects automatically necessary orbitals to join the interaction with attacking monomer in "orbital basis concept" using Eq. 6.4 even if the B-region is small, and then we can start it with small starting cluster without loss of accuracy.

To confirm the reliability of our treatment for larger basis sets including diffuse and polarization functions, we perform the same calculation for lycopene using HF/6-31+G(d,p) basis set. From Table 6.8, one can see that the obtained total energy errors using large basis set are similar to the above results at 3-21G level or even more improved. Also we elongate it only with B region = 2 units but elongation starts from end of 7th unit after the calculations using all the frozen and active LMOs for 4, 5 and 6 units because the condition Eq. 6.3 was not satisfied within the threshold until the 7th unit. This calculation with small B region size is actually the same as that using Bregion = 5 units. In principle, B region that is large enough to satisfy the required accuracy should be defined before starting elongation calculations. Nevertheless, in the new elongation method, the elongation is automatically initiated regardless of B region size, which always makes the task to define B region being easy even if starting B region = 1. These excellent results suggest no basis set dependency as well as no B region size dependency on the accuracy of our new treatment.

Table 6.9 Error per atom (in a.u.) introduced by the new elongation method for $C_{60}-(C_4\ H_4)_n-H_2TPP$ system at the HF/3−21G level

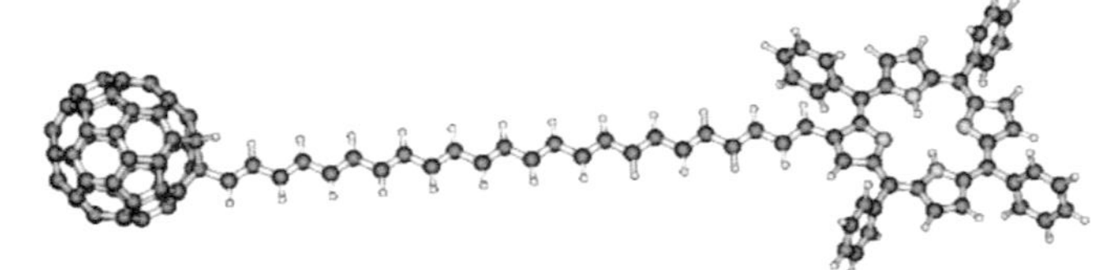

Dyad	Number of atoms	Total energy Conv (in a.u.)	Total energy Elg (in a.u.)	Δ E (Elg−Conv)	New_Elg Error/atom
$C_{60}-C_4H_4-$H2TPP	146	−4302.7457358264	−4302.7457358232	3.17E − 09	2.17E − 11
$C_{60}-(C_4H_4)_2-$H2TPP	154	−4455.6587183241	−4455.6587162673	2.06E − 06	1.34E − 08
$C_{60}-(C_4H_4)_3-$H2TPP	162	−4608.5726760087	−4608.5726728645	3.14E − 06	1.94E − 08
$C_{60}-(C_4H_4)_4-$H2TPP	170	−4761.4867339248	−4761.4867312944	2.63E − 06	1.55E − 08
$C_{60}-(C_4H_4)_5-$H2TPP	178	−4914.4008358177	−4914.4007676444	6.82E − 05	3.83E − 07
$C_{60}-(C_4H_4)_6-$H2TPP	186	−5067.3149696281	−5067.3149674217	2.21E − 06	1.19E − 08

6.5.5 Fullerene–Polyacethylene–H_2TPP

Next target of application is the examination of our treatment on some large intermolecular complexes through nano-wire of conducting oligomer. Recently fullerene and porphyrin dyads of donor–acceptor (D–A) character are subjects of considerable interest because those two gigantic substitutions in the terminal are attracted to form host-guest complex. To understand the nature of the interaction between fullerene and chromophore dyads like pophyrin with different electron-donor character, several theoretical studies have been carried out for those complexes [48–51]. Here we adopted one of the most studied free base tertaphenylporphyrin (H_2TPP) that constitutes an important class of π conjugated organic chromophores and used to functionalize nanowires [52–56].

First, we applied our revised elongation method to the following C_{60}-$(C_4H_4)_n$-H_2TPP system to see if this method is applicable or not to such gigantic delocalized systems. The calculations were performed by increasing the number of the (C_4H_4) units in the central wire part keeping terminated by C_{60} and H_2TPP. The Table 6.9 shows that the accuracy is satisfied for such delocalized large systems also though it has lost around by one-order compared to that of other examples presented in aforementioned subsections.

6.5.6 Fullerene–Oligo(2,5-thienylene-ethynylene)–H_2TPP

Furthermore, we extend the system to a more realistic nanowire complex shown below in which the polyacetylene of above example was replaced by head-to-tail coupled polythiophene derivatives—[oligo(2,5-thienylene-ethynylene)-](OTE)-terminated by the same fullerene and H_2TPP. The OTE wire part is a class of conjugated oligomers with a high shape-persistence as rigid rods and special attention is focused on the systems with terminal donor–acceptor substitution because of a strong push-pull effect [57–59].

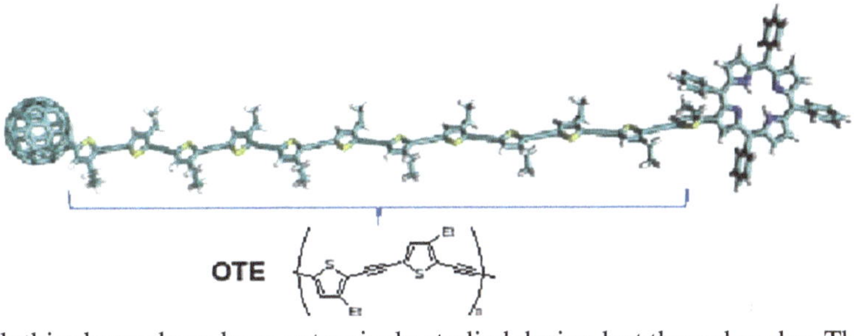

Polythiophenes have been extensively studied during last three decades. The pioneering works of Alan Heeger, Alan MacDiarmid, and Hideki Shirakawa in the field of conducting polymers gained international recognition by the awarding of the 2000 Nobel Prize in Chemistry "for the discovery and development of conductive polymers" [60–62]. These materials among other notable properties exhibit quite high electrical conductivity, which originates from the delocalization of electrons along the polymer backbone. Because of the extreme conductivity properties they are often regarded as "synthetic metals". The electron delocalization gives rise also to optical phenomena. The optical properties of these materials respond to external perturbations, with significant color shifts in response to changes in solvent environment, temperature, applied potential, and binding to other molecules. The color and conductivity effects are governed by common mechanism which relies on twisting of the polymer backbone, disrupting conjugation. This makes conjugated polymers attractive for manufacturing sensors that can provide a wide range of optical and electronic responses. For this reason, the applicability of our method to such delocalized nanowires should be carefully checked for further applications of this method toward designing functional nanowire systems.

If our aim is only the final systems of the nanowire with terminal donor and acceptor, we can elongate either from C_{60} terminal of the wire and terminate with Porphyrin or from the central to the both end part as already performed for push-pull system [39]. Before calculation of the whole system, the accuracy of our treatment was investigated only for OTE wire part and the errors obtained became improved by around three orders of magnitude in total energy/atom compared to those by the older conventional elongation method as shown in the Table 6.10, giving extremely small errors $\sim 10^{-12}$ a.u./atom. So, we confirmed that the orbital shift treatment is very useful method to get more accurate results for such π-electron delocalized systems. Additionally, as one of final purposes we calculated (hyper) polarizability of the OTE wire system and the obtained α, β and γ values by the new elongation method and the conventional method were listed below the table for energy comparison. From very sensitive numerical instability in higher order derivatives by electronic field of total energy, the error in γ value becomes larger than α and β, but still is within $\sim 10^{-2}$ % of the total value. These accuracies are very promising to investigate NLO properties even in such strongly delocalized systems. These NLO results from FF method depend only on the accuracy of the total energy for different fields and thus the obtained accuracies are reasonable as we already published NLO results for these field strengths (see for example, Refs. [37–42] for NLO results without the orbital shifted method). The NLO investigations for other systems are systematically ongoing and will be reported soon, though in the present article we just demonstrate one example to show how our new treatment is working for this property.

It's also very useful to see the wire's effect under the influence of the big terminal substitutions. As an another way to investigate the relationship between wire length and physical property, we can elongate the wire part -OTE- as the terminal substitution is fixed as shown in Table 6.11. We can elongate the chain with terminated H_2TPP molecules instead of hydrogen. This way for elongation, however, is time consuming as we have to keep the two big terminal molecules during elongation at least until AO-cut starts for removing C_{60}. Therefore, the right end of the chain was elongated with terminal hydrogen and then H_2TPP was attached only at the final step, by which we can save computational time during elongation if we focus only on the final C_{60}-[OTE]-H_2TPP with different OTE lengths. The three models with different

Table 6.10 Error per atom (in a.u.) introduced by the old and new elongation methods for OTE wire system at the HF/3-21G level and Dipole Moment and (Hyper)polarizability comparison at $n = 20$

Number of Unit*	Total energy Conv (in a.u.)	Old_Elg Error/atom	New_Elg Error/atom
12	−8276.7158675228	6.00E − 10	3.33E − 12
14	−9668.4393841348	2.00E − 10	9.53E − 13
16	−11060.1629006892	1.30E − 09	5.42E − 12
18	−12451.8864172189	1.00E − 09	3.71E − 12
20	−13843.6253615705	1.40E − 09	4.67E − 12
Method	Dipole Moment & (Hyper)Polarizability		Error (%)
New_ Elg	μ_z(Debye)	**1.648**	
Conv		**1.648**	6.68E − 06
New_Elg	α_{zz}(a.u.)	**337.50**	
Conv		**337.50**	2.73E − 05
New_Elg	γ_{zzzz}(a.u.)	**−1804**	
Conv		**−1797**	4.03E − 01

*Unit = two thiophene rings

Table 6.11 Error per atom (in a.u.) introduced by the new elongation method for C_{60}-[OTE]-H_2TPP system at the HF/3-21G level, where only final energies for [1], [2] and [3] were listed

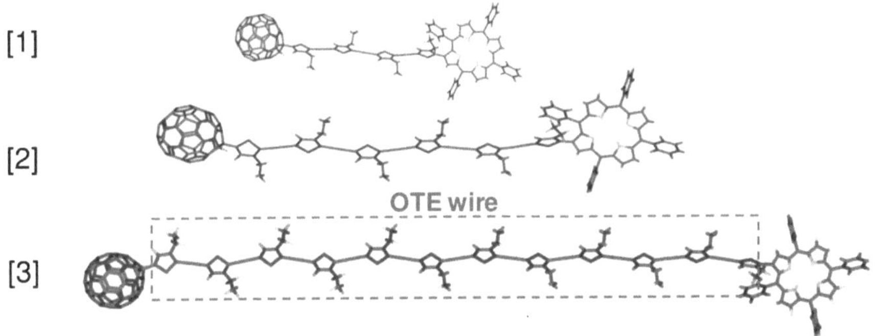

Dyad	Number of atoms	Total energy Cony (in a.u.)	Total energy Elg (in a.u.)	ΔE(Elg−Conv)	New_Elg Error/atom
[1]	196	−6875.4797176171	−6875.479717505	1.12E − 07	5.72E − 10
[2]	226	−8275.9355460483	−8275.9355460401	8.20E − 09	3.63E − 11
[3]	316	−12477.3014981261	−12477.301498125	1.10E − 09	3.48E − 12

OTE lengths show the final structures after elongation. The calculated total energies show the excellent agreement with those by the direct calculations for whole system using the conventional direct method. These applications are promising enough to provide very reliable NLO property for gigantic systems that cannot be treated using the conventional direct method.

6.6 Conclusions

The elongation method has been further improved by orbital selection treatment to reconstruct necessary active RLMOs. By this correction technique, a few initially defined frozen orbitals that are not well localized into frozen region are shifted to the active space. After testing it on SW-BN/C nanotube, polyacene, β-carotene, lycopene, fullerene-C_4H_4-H_2TPP, OTE, and fullerene- OTE-H_2TPP we have confirmed that the improved method becomes more applicable to strongly delocalized systems with high accuracy and with even more efficiency. This work now allows one to perform NLO property simulations for large systems with delocalization. In addition to applications, further development is ongoing to treat systems which require multireference ground state wave functions, that is RASSCF method and CASPT2 to treat dynamic correlation. We look forward in the near future to present new results in this area.

References

1. Imamura, A., Aoki, Y., Maekawa, K.: A theoretical synthesis of polymers by using uniform localization of molecular orbitals: proposal of an elongation method. J. Chem. Phys. 95, 5419–5431 (1991)
2. Aoki, Y., Imamura, A.: Local density of states of aperiodic polymers using the localized orbitals from an ab initio elongation method. J. Chem. Phys. **97**, 8432–8440 (1992)
3. Aoki, Y., Suhai, S., Imamura, A.: An efficient cluster elongation method in density functional theory and its application to poly-hydrogen-bonding molecules. J. Chem. Phys. **101**, 10808–10823 (1994)
4. Makowski, M., Korchowiec, J., Gu, F.L., Aoki, Y.: Describing electron correlation effects in the framework of the elongation method -Elongation-MP2: formalism, implementation and efficiency. J. Comput. Chem. **31**, 1733–1740 (2010)
5. Gu, F.L., Imamura, A., Aoki, Y.: Elongation method for polymers and its application to nonlinear optics, in atoms, molecules and clusters in electric fields: theoretical approaches to the calculation of electric polarizabilities. In: Maroulis, G. (ed.) Computational Numerical and Mathematical Methods in Sciences and Engineering, vol. 1, pp. 97–177. Imperial College Press, London (2006)
6. Goedecker, S.: Linear scaling electronic structure methods. Rev. Mod. Phys. **71**, 1085–1123 (1999)
7. Salek, P., Hoest, S., Thoegersen, L., Joergensen, P., Manninen, P., Olsen, J., Jansik, B., Reine, S., Pawlowski, F., Tellgren, E., Helgaker, T., Coriani, S.: Linear-scaling implementation of molecular electronic self-consistent field theory. J. Chem. Phys. **126**, 114110 (2007)

8. Goedecker, S., Teter, M.: Tight-binding electronic-structure calculations and tight-binding molecular dynamics with localized orbitals. Phys. Rev. B **51**, 9455–9464 (1995)
9. Stephan, U., Drabold, D.: Order-N projection method for first-principles computations of electronic quantities and Wannier functions. Phys. Rev. B **57**, 6391–6407 (1998)
10. Li, X.-P., Nunes, W., Vanderbilt, D.: Density-matrix electronic-structure method with linear system-size scaling. Phys. Rev. B **47**, 10891–10894 (1993)
11. Zhao, Q., Yang, W.: Analytical energy gradients and geometry optimization in the divide-and-conquer method for large molecules. J. Chem. Phys. **102**, 9598–9603 (1995)
12. Kim, J., Mauri, F., Galli, G.: Total-energy global optimizations using nonorthogonal localized orbitals. Phys. Rev. B **52**, 1640–1648 (1995)
13. Hernandez, E., Gillan, M.: Self-consistent first-principles technique with linear scaling. Phys. Rev. B **51**, 10157–10160 (1995)
14. Yang, W.: Direct calculation of electron density in density-functional theory. Phys. Rev. Lett. **66**, 1438–1441 (1991)
15. Yang, W., Lee, T.-S.: A density-matrix divide-and-conquer approach for electronic structure calculations of large molecules. J. Chem. Phys. **103**, 5674–5678 (1995)
16. Akama, T., Kobayashi, M., Nakai, H.: Implementation of divide-and-conquer method including Hartree-Fock exchange interaction. J. Comput. Chem. **28**, 2003–2012 (2007)
17. Kobayashi, M., Imamura, Y., Nakai, H.: Alternative linear-scaling methodology for the second-order Møller-Plesset perturbation calculation based on the divide-and-conquer method. J. Chem. Phys. **127**, 074103 (2007)
18. Kitaura, K., Ikeo, E., Asada, T., Nakano, T., Uebayasi, M.: Fragment molecular orbital method: an approximate computational method for large molecules. Chem. Phys. Lett. **313**, 701–706 (1999)
19. Fedorov, D.G., Kitaura, K.: The importance of three-body terms in the fragment molecular orbital method. J. Chem. Phys. **120**, 6832–6840 (2004)
20. Fedorov, D.G., Kitaura, K.: On the accuracy of the 3-body fragment molecular orbital method (FMO) applied to density functional theory. Chem. Phys. Lett. **389**, 129–134 (2004)
21. Fedorov, D.G., Kitaura, K.: Extending the power of quantum chemistry to large systems with the fragment molecular orbital method. J. Phys. Chem. A **111**, 6904–6914 (2007)
22. Mochizuki, Y., Koikegami, S., Nakano, T., Amari, S., Kitaura, K.: Large scale MP2 calculations with fragment molecular orbital scheme. Chem. Phys. Lett. **396**, 473–479 (2004)
23. Mochizuki, Y., Fukuzawa, K., Kato, A., Tanaka, S., Kitaura, K., Nakano, T.: A configuration analysis for fragment interaction. Chem. Phys. Lett. **410**, 247–253 (2005)
24. Mochizuki, Y., Ishikawa, T., Tanaka, K., Tokiwa, H., Nakano, T., Tanaka, S.: Dynamic polarizability calculation with fragment molecular orbital scheme. Chem. Phys. Lett. **418**, 418–422 (2006)
25. Mochizuki, Y., Yamashita, K., Murase, T., Nakano, T., Fukuzawa, K.: Takematsu, K., Watanabe, H., Tanaka, S.: Large scale FMO-MP2 calculations on a massively parallel-vector computer. Chem. Phys. Lett. **457**, 396–403 (2008)
26. White, S.R.: Density matrix formulation for quantum renormalization groups. Phys. Rev. Lett. **69**, 2863–2866 (1992)
27. Kurashige, Y., Yanai, T.: High-performance ab initio density matrix renormalization group method: applicability to large-scale multireference problems for metal compounds. J. Chem. Phys. **130**, 234114 (2009)
28. Mizukami, W., Kurashige, Y., Yanai, T.: Communication: novel quantum states of electron spins in polycarbenes from ab initio density matrix renormalization group calculations. J. Chem. Phys. **133**, 091101 (2010)
29. Gu, F.L., Aoki, Y., Korchowiec, J., Imamura, A., Kirtman, B.: A new localization scheme for the elongation method. J. Chem. Phys. **121**, 10385–10391 (2004)
30. Aoki Y, Gu FL (2009) Generalized elongation method: from one-dimension to three-dimension. In: Champagne, B., Gu, F.L., Luis, J.M., Springborg M (org) International Conference of Computational Methods in Sciences and Engineering 2009 (ICCMSE 2009), pp. 46–49

31. Makowski, M., Gu, F.L., Aoki, Y.: Elongation-CIS method: describing excited states of large molecular systems in regionally localized molecular orbital basis. J. Comput. Meth. Sci. Eng. **10**, 473–481 (2010)
32. Korchowiec, J., Gu, F.L., Imamura, A., Kirtman, B., Aoki, Y.: Elongation method with cutoff technique for linear SCF scaling. Int. J. Quantum. Chem. **102**, 785–794 (2005)
33. Makowski, M., Korchowiec, J., Gu, F.L., Aoki, Y.: Efficiency and accuracy of the elongation method as applied to the electronic structures of large systems. J. Comp. Chem. **27**, 1603–1619 (2006)
34. Korchowiec, J., Lewandowski, J., Makowski, M., Gu, F.L., Aoki, Y.: Elongation cutoff technique armed with quantum fast multipole method for linear scaling. J. Comput. Chem. **30**, 2515–2525 (2009)
35. Korchowiec, J., Silva, P., Makowski, M., Gu, F.L., Aoki, Y.: Elongation cutoff technique at Kohn-Sham level of theory. Int. J. Quantum. Chem. **110**, 2130–2139 (2010)
36. Gu, F.L., Aoki, Y., Imamura, A., Bishop, D.M., Kirtman, B.: Application of the elongation method to nonlinear optical properties: finite field approach for calculating static electric (hyper)polarizabilities. Mol. Phys. **101**, 1487–1494 (2003)
37. Ohnishi, S., Gu, F.L., Naka, K., Imamura, A., Kirtman, B., Aoki, Y.: Calculation of static (hyper)polarizabilities for π-conjugated donor/acceptor molecules and block copolymers by the elongation finite-field method. J. Phys. Chem. A **108**, 8478–8484 (2004)
38. Gu, F.L., Guillaume, M., Botek, E., Champagne, B., Castet, F., Ducasse, L., Aoki, Y.: Elongation method and supermolecule approach for the calculation of nonlinear susceptibilities. Application to the 3-methyl-4-nitropyridine 1-oxide and 2-methyl-4-nitroaniline crystals. J. Comput. Meth. Sci. Eng. 6, 171–188 (2006)
39. Ohnishi, S., Orimoto, Y., Gu, F.L., Aoki, Y.: Nonlinear optical properties of polydiacetylene with donor-acceptor substitution block. J. Chem. Phys. **127**, 084702 (2007)
40. Chen, W., Yu, G.-T., Gu, F.L., Aoki, Y.: Investigation on the electronic structures and nonlinear optical properties of pristine boron nitride and boron nitride-carbon heterostructured single-wall nanotubes by the elongation method. J. Phys. Chem. C **113**, 8447–8454 (2009)
41. Yan, L.K., Pomogaeva, A., Gu, F.L., Aoki, Y.: Theoretical study on nonlinear optical properties of metalloporphyrin using elongation method. Theor. Chem. Acc. **125**, 511–520 (2010)
42. Schmidt, M.W., Baldridge, K.K., Boatz, J.A., Elbert, S.T., Gordon, M.S.: Jensen, J.H., Koseki, S., Matsunaga, N., Nguyen, K.A., Su, S.J., Windus, T.L., Dupuis, M., Montgomery, J.A.: General atomic and molecular electronic structure system. J. Comput. Chem. **14**, 1347–1363 (1993)
43. Pomogaeva, A., Gu, F.L., Imamura, A., Aoki, Y.: Electronic structures and nonlinear optical properties of supramolecular associations of benzo-2,1,3-chalcogendiazoles by the elongation method. Theor. Chem. Acc. **125**, 453–460 (2010)
44. Blase, X., Charlier, J.C., De Vita, A., Car, R.: Structural and electronic properties of composite BxCyNz nanotubes and heterojunctions. Appl. Phys. A **68**, 293–300 (1999)
45. Terrones, M., Romo-Herrera, J.M., Cruz-Silva, E., Lopez-Urias, F., Munoz-Sandoval, E., Velazquez-Salazar, J.J., Terrones, H., Bando, Y., Golberg, D.: Pure and doped boron nitride nanotubes. Mater. Today **10**, 30–38 (2007)
46. Gaussian 03, Revision C.02, Frisch M.J., Trucks, G.W., Schlegel, H.B., Scuseria, G.E., Robb, M.A., Cheeseman, J.R., Montgomery, J.A. Jr, Vreven, T., Kudin, K.N., Burant, J.C., Millam, J.M., Iyengar, S.S., Tomasi, J., Barone, V., Mennucci, B., Cossi, M., Scalmani, G., Rega, N., Petersson, G.A., Nakatsuji, H., Hada, M., Ehara, M., Toyota, K., Fukuda, R., Hasegawa, J., Ishida, M., Nakajima, T., Honda, Y., Kitao, O., Nakai, H., Klene, M., Li, X., Knox, J.E., Hratchian, H.P., Cross, J.B., Adamo, C., Jaramillo, J., Gomperts, R., Stratmann, R.E., Yazyev, O., Austin, A.J., Cammi, R., Pomelli, C., Ochterski, J.W., Ayala, P.Y., Morokuma, K., Voth, G.A., Salvador, P., Dannenberg, J.J., Zakrzewski, V.G., Dapprich, S., Daniels, A.D., Strain, M.C., Farkas, O., Malick, D.K., Rabuck, A.D., Raghavachari, K., Foresman, J.B., Ortiz, J.V., Cui, Q., Baboul, A.G., Clifford, S., Cioslowski, J., Stefanov, B.B., Liu, G., Liashenko, A., Piskorz, P., Komaromi, I., Martin, R.L., Fox, D.J., Keith, T., Al-Laham M.A., Peng, C.Y., Nanayakkara, A., Challacombe, M., Gill, P.M.W., Johnson, B., Chen, W., Wong, M.W., Gonzalez, C., Pople, J.A., Gaussian, Inc., Wallingford, CT (2004)

47. http://xray.bmc.uu.se/hicup/LYC/index.html#COO
48. Liao, M.-S., Watts, J.D., Huang, M.-J.: Dispersion-corrected DFT calculations on C60-porphyrin complexes. Phys. Chem. Chem. Phys. **11**, 4365–4374 (2009)
49. Loboda, O., Zalesny, R., Avramopoulos, A., Papadopoulos, M.G., Artacho, E.: Linear-scaling calculations of linear and nonlinear optical properties of [60]fullerene derivatives. In: Maroulis, G., Simos, T.E. (eds.) Computational Methods in Sciences and Engineering. Advances In Computational Science Book Series: AIP Conference Proceedings, vol 1108, 198–204 (2009)
50. Loboda, O., Zalesny, R., Avramopoulos, A., Luis, J.-M., Kirtman, B., Tagmatarchis, N., Reis, H., Papadopoulos, M.G.: Linear and nonlinear optical properties of [60]fullerene derivatives. J. Phys. Chem. A **113**, 1159–1170 (2009)
51. Zalesny, R., Loboda, O., Iliopoulos, K., Chatzikyriakos, G., Couris, S., Rotas, G., Tagmatarchis, N., Avramopoulose, A., Papadopoulos, M.G.: Linear and nonlinear optical properties of triphenylamine-functionalized C60: insights from theory and experiment. Phys. Chem. Chem. Phys. **12**, 373–381 (2010)
52. Palummo, M., Hogan, C., Sottile, F., BagalÃ, P., Rubio, A.: Ab initio electronic and optical spectra of free-base porphyrins: the role of electronic correlation. J. Chem. Phys. **131**, 084102 (2009)
53. Yeon, K.Y., Jeong, D., Kim, S.K.: Intrinsic lifetimes of the Soret bands of the free-base tetraphenylporphine (H2TPP) and Cu(II)TPP in the condensed phase. Chem. Commun. **46**, 5572–5574 (2010)
54. Li, C., Ly, J., Lei, B., Fan, W., Zhang, D., Han, J., Mayyappan, M., Thompson, C., Zhou, M.: Data storage studies on nanowire transistors with self-assembled porphyrin molecules. J. Phys. Chem. B **108**, 9646–9649 (2004)
55. Kwok, K.S.: Materials for future electronics. Mater. Today **6**, 20–27 (2003)
56. Duan, X., Huang, Y., Lieber, C.M.: Nonvolatile memory and programmable logic from molecule-gated nanowires. Nano. Lett. **2**, 487–490 (2002)
57. Meier H, Mühling, B.: Synthesis and properties of oligo(2,5-thienylene)s. In: Waring, A. (ed.) ARKIVOC: Special Issue "5th Eurasian Conference on Heterocyclic Chemistry" ix:57–69 (2009)
58. Obara, Y., Takimiya, K., Aso, Y., Otsubo, T.: Synthesis and photophysical properties of [60]fullerene-oligo(thienylene-ethynylene) dyads. Tetrahedron Lett. **42**, 6877–6881 (2001)
59. Tormos, G.V., Nugara, P.N., Lakshmikantham, M.V., Gava, M.P.: Poly(2,5-thienylene ethynylene) and related oligomers. Synth. Met. **53**, 271–281 (1993)
60. Shirakawa, H., Louis, E.J., MacDiarmid, A.G., Chiang, C.K., Heeger, A.J.: Synthesis of electrically conducting organic polymers: halogen derivatives of polyacetylene, (CH)x. J. Chem. Soc. Chem. Commun. **474**, 578–580 (1977)
61. Chiang, C.K., Druy, M.A., Gau, S.C., Heeger, A.J., Louis, E.J., MacDiarmid, A.G., Park, Y.W., Shirakawa, H.: Synthesis of highly conducting films of derivatives of polyacetylene, (CH)x. J. Am. Chem. Soc. **100**, 1013–1015 (1978)
62. MacDiarmid, A.G.: Synthetic metals: a novel role for organic polymers. Synth. Metals **125**, 11–22 (2001)